W9-ABN-163

Boundary methods
An algebraic theory

Boundary Methods

An Algebraic Theory

Ismael Herrera
Institute of Geophysics
National University of Mexico

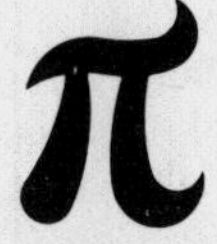

Pitman Advanced Publishing Program
BOSTON · LONDON · MELBOURNE

PITMAN PUBLISHING LIMITED
128 Long Acre, London WC2E 9AN

PITMAN PUBLISHING INC.
1020 Plain Street, Marshfield, Massachusetts 02050

Associated Companies
Pitman Publishing Pty Ltd, Melbourne
Pitman Publishing New Zealand Ltd, Wellington
Copp Clark Pitman, Toronto

First published 1984

AMS Subject Classifications: Main 35A25, 35A35, 35A40, 35C10
Subsidiary 35G05, 35G10, 35G15

Library of Congress Cataloging in Publication Data

Herrera, Ismael.
Boundary methods.

(Applicable mathematics series)
Bibliography: p.
Includes index.
1. Boundary value problems—Numerical solutions.
2. Approximation theory. I. Title. II. Series.
QA379.H47 1984 515.3′5 84-3209
ISBN 0-273-08635-9

British Library Cataloguing in Publication Data

Herrera, Ismael.
Boundary methods.—(Applicable mathematics series)
1. Boundary value problems
I. Title II. Series
515.3′5 QA379

ISBN 0-273-08635-9

Filmset and printed in Northern Ireland by The Universities Press (Belfast) Ltd., and bound at the Pitman Press, Bath, Avon.

Contents

Preface vii

1 Introduction 1

Part 1 Algebraic theory

2 Preliminary notions and notations 13

3 Formal adjoints and Green's formulas 19

4 Abstract characterization of Green's formulas. Regular subspaces and canonical decompositions 26

5 Green's formulas for operators defined in discontinuous fields 39

6 Illustrations of Green's formulas 49

7 Illustrations of jump operators 60

Part 2 Boundary methods

8 Scope 69

9 The subspace I_P 82

10 Immersion in a Hilbert space 89

11 Criterion of completeness 98

12 Solution of boundary value problems 125

References 131

DEDICATED TO

Roberto Vázquez
from whom I learned deep mathematics,

Richard E. Meyer
from whom I learned applied mathematics, and

Emilio Rosenblueth
from whom I learned to apply mathematics.

Preface

This book is designed to provide an integrated presentation of an abstract theory of boundary value problems which I have developed over several years of work. My interest in forming an abstract theory started in 1973 in connection with variational principles. Previously, in the early 1960s, I had published, mainly in the *Bulletin of the Seismological Society of America* and the *Journal of Geophysical Research*, work on diffraction problems that occur in geophysics and seismic engineering. The abstract theory was greatly enriched when it was brought into contact with this more concrete area of interest. A breakthrough in the development of the theory, which was published in a sequence of three papers [21, 22, 23], was achieved when trying to incorporate in my general framework results obtained by Mei and Chen for linearized free surface flows.

Further growth of the theory was stimulated by developments in boundary element methods, a subject that has received much attention in the engineering literature, to a large extent due to the activity of Brebbia. A little later, it was possible to incorporate my earlier work on biorthogonal Fourier series in the setting supplied by the algebraic theory. This renewed interest in the subject was prompted by work that has been done in recent years by Joseph, Spence and Gregory.

Of the three kinds of applications that have been mentioned (variational principles, boundary methods and biorthogonal Fourier series), only Trefftz procedure is discussed in this book. The application to variational principles is well developed and a preliminary survey is given in [31]. A discussion of biorthogonal systems in the setting of the algebraic theory was published in [37] and a more complete discussion may be found in [35].

I want to express my appreciation to those colleagues who collaborated with me at different periods during the development of the theory: namely, Jacobo Bielak (Carnegie-Mellon University), Michael J. Sewell (Reading University), Roland England,

Hervé Gourgeon, David A. Spence (Imperial College), Francisco Sánchez-Sesma, Javier Avilés and Gonzalo Alduncin; to Ben Noble (University of Wisconsin-Madison), whose work and interest in variational principles prompted my own; to J. Tinsley Oden (University of Texas at Austin), Jacques L. Lions (Institut National de Recherche en Informatique et en Automatique) and Olek C. Zienkiewicz (Swansea University), for useful discussions; and to George F. Pinder, whose enthusiasm and example have encouraged me to finish the book. Special mention is made of valuable contributions made by John Evans (University of California at San Diego), and also Dietrich Ullenbrock (University of Wisconsin-Madison), who brought to my attention the work on symplectic geometry. M. Zuhair Nashed (University of Delaware), Sung J. Lee (University of Florida) and Earl A. Coddington (UCLA) supplied me with extensive information on their work. Also, some valuable suggestions were made by Alfonso Vignoli (University of Rome). Finally, I am grateful for the assistance of Mrs Martha Cerrilla de Canudas, who typed the manuscript.

March, 1984 I.H.

1 Introduction

Abstract formulations of boundary value problems have long been available. A classical approach is presented, for example, in Dunford and Schwartz [1]. A different point of view, based on the notions of adjoint subspaces introduced by Arens [2], has been developed by Coddington [3–5] and more recently by Lee [6–12]. Apparently, it has been applied exclusively to ordinary differential equations.

In a sequence of papers [13–44] the author has introduced an abstract formulation which is quite suitable for partial differential equations.† This theory is based on an algebraic structure which systematically occurs in boundary value problems which are linear. This structure is interesting because of its simplicity and beauty. Also, it has relevant applications. Thus far, the latter have been along three main lines:

(1) variational principles;
(2) numerical solution of boundary value problems; and
(3) development of biorthogonal systems of functions, to obtain generalized Fourier series developments.

The algebraic theory is formulated in the setting of general functional-valued operators defined on arbitrary linear spaces which, generally, do not possess an inner product or metric [15]. This supplies greater flexibility. The notions of boundary operator, formal adjoint and formal symmetry are defined abstractly for functional-valued operators. This allows the introduction of abstract Green's formulas.

Although there is no basic reason for restricting attention to formally symmetric operators, and work is being carried out to extend it to general nonsymmetric operators, at present the theory is well developed only for the former class of operators. A skew-

† By invitation, this theory was presented at a special session on Green's formulas and abstract adjoints at the 23rd Annual Meeting of the American Mathematical Society, at Denver, Colorado, in January, 1983.

symmetric bilinear form is associated with every formally symmetric operator and it is shown that there is a one-to-one correspondence between abstract Green's formulas and canonical decompositions of the space. By 'canonical decomposition' is meant a pair of regular isotropic subspaces which span the whole linear space in which the operators are defined. The regular isotropic subspaces occurring in a canonical decomposition are shown to be completely regular (i.e. maximal isotropic) necessarily. The algebraic structure which is obtained in this manner is closely related to symplectic geometry (see, for example, Abraham and Marsden [45] and Weinstein [46]). In particular, the following relations can be established:

(i) regular subspaces are isotropic;
(ii) Lagrangean subspaces are completely regular (i.e. maximal isotropic); and
(iii) every Lagrangean splitting is a canonical decomposition.

The main differences from symplectic geometry derive from the fact that the basic skew-symmetric bilinear forms considered by the author are degenerate. This is essential in applications to boundary value problems.

The relations between the author's theory and that of Arens, Coddington and Lee are less clear. In recent work to be published, the author has shown that the model of the latter approach can be incorporated in the general framework of the algebraic theory. However, the possible implications of this must be the subject of further research.

A characterization of Green's formulas by means of linear subspaces supplies considerable flexibility to the theory. It allows, for example, the development of an abstract formulation of problems in discontinuous fields, with prescribed jump conditions. This has been developed under the general heading of the problem of matching or connecting. Corresponding abstract Green's formulas were given introducing 'jump operators' [31].

When classical approaches are used, the heat operator cannot be associated with a formally symmetric operator. Within the framework of the algebraic theory this is possible. Also, classical Green's formulas for the wave equation are not suitable for the formulation of well-posed problems. The same is true for dynamic problems: for example, in elasticity. This is reflected in the fact that variational principles for time-dependent problems were not developed until 1964, when Gurtin [47, 48] reported the first ones. His approach and principles can be simplified (see, for example, [14]). Essentially, what is required is to develop Green's formulas that are suitable for well-posed problems. This is achieved, however, by means of Gurtin's convolutions.

As these examples illustrate, the theory is applicable irrespective of the type of equation, or system of equations; in particular, it can be applied to both time-independent and time-dependent equations. In this respect, the extension to operators that are not required to be formally symmetric would lead to a theory applicable to arbitrary partial differential equations or systems of such equations.

Variational principles have always played an important role in mathematical physics [49]. However, a revival of interest in the subject arose from the development of numerical methods for use with computers in the handling of partial differential equations [50, 51].

The application of the algebraic theory to the formulation of variational principle is straightforward. Indeed, an abstract Green's formula supplies an expression of the form

$$P - P^* = B - B^*, \tag{1.1}$$

where the asterisk denotes the adjoint or transpose of the operator. Also, B and B^* are boundary operators. Equation (1.1) implies

$$P - B = P^* - B^* = (P - B)^*. \tag{1.2}$$

This exhibits $P - B$ as a symmetric operator and the formulation of a variational principle for a boundary value problem, in which Pu together with the boundary values Bu, are prescribed, is immediate. For time-dependent problems, initial conditions can be incorporated in B, if the problem is formulated in a space–time region.

Variational principles for problems formulated in discontinuous fields and with prescribed jump conditions are useful in numerical applications. They have been developed in an *ad hoc* manner for some specific applications. For elasticity they are discussed by Prager [52] and Nemat-Nasser [53, 54].

A related class of principles that has received attention recently is used to match a region which is discretized with one which is treated numerically [55, 56]. Diffraction problems constitute a

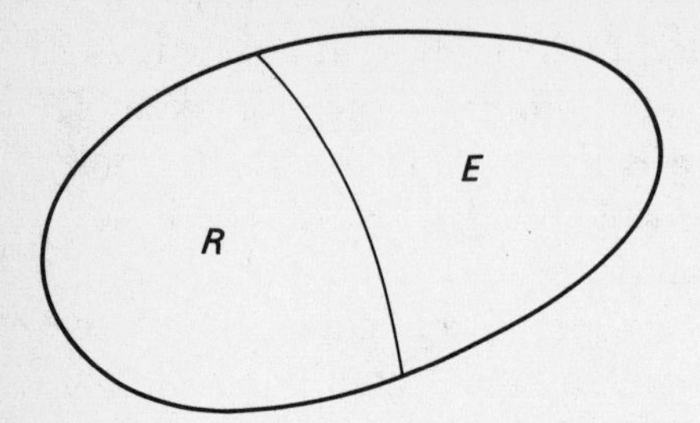

Fig. 1.1 Regions for the problem of matching

typical example [29, 30, 40, 55] for which the region treated analytically is unbounded. In this case one has to formulate variational principles which are subjected to restrictions of continuation type; by this it is meant that the numerical solution is required to be such that it matches smoothly with an analytical solution of the differential equation across the boundary which is common to both regions (Fig. 1.1). General results about these kinds of principles are given in [21–24] and a more systematic discussion appears in [31]. For problems with prescribed jumps they can be derived using the abstract Green's formulas for the problem of matching or connecting, mentioned previously. When the problems are subjected to restrictions of continuation type, the arguments require slight modification [31].

Another type of application is boundary methods. A boundary method is nowadays usually taken to mean a numerical procedure in which a subregion, or the entire region, may be left out of the numerical treatment by the use of available analytical solutions (or, more generally, previously computed solutions). Boundary methods, by reducing the dimensions involved in the problems, lead to considerable economy in the numerical work and constitute a very convenient manner of treating unbounded regions adequately by numerical means. Generally, the dimensionality of the problem is reduced by one; even when part of the region is treated by finite elements, the size of the discretized domain is reduced [56, 57].

There are two main approaches for the formulation of boundary methods: one is based on the use of boundary integral equations, and the other on the use of complete systems of solutions. In numerical applications, the first method has received most of the attention [58, 59]. This is despite the fact that the use of complete systems of solutions presents important numerical advantages: e.g., it avoids the introduction of singular integral equations and it does not require the construction of a fundamental solution. The latter point is especially relevant in connection with complicated problems, for which it may be extremely laborious to build up a fundamental solution. This is illustrated by the fact that there are methods for synthesizing fundamental solutions starting from plane waves, which in Chapter 11 is shown to be a complete system [40].

One may advance some possible explanations for this situation. Although the use of complete families of solutions to solve boundary value and initial-boundary value problems in partial differential equations dates back to the time of Fourier, many of its applications have been based on the method of separation of variables; this has led to the common, but false, belief that complete systems of solutions have to be constructed specifically for a given region. Actually, this is not the case; frequently systems of solutions are complete independently of the detailed shape of the region considered. The theory of integral operators and their use in constructing complete families of solutions to partial differential equations with analytic coefficients [60–65] is well known. Usually, such methods yield systems which are complete for general classes of regions.

The use of complete systems of solutions is frequently associated with the name of Trefftz [66]. The idea of his method [67] consists in looking for approximate solutions among the appropriate class of functions that exactly satisfy the differential equation but do not necessarily satisfy the prescribed boundary conditions. Although Trefftz's original formulation was linked to a variational principle, this is not a requirement. Indeed, complete systems of solutions can be used to treat differential equations which are linear but otherwise arbitrary (see, for example, [38]). For parabolic equations Colton has proposed a numerical procedure to fit the boundary and initial data [68]. Rosenblueth has explored some alternatives [69].

The method has been used in many fields. For example, applications to Laplace's equation are given by Mikhlin [70], to the biharmonic equation by Rektorys [71] and to elasticity by Kupradze *et al.* [72]. Also many scattered contributions to the method can be found in the literature. Special mention is made here of work by Amerio, Fichera, Kupradze, Picone and Vekua [73–76, 64]. For parabolic equations Colton has constructed families which are complete in regions which are to a large extent arbitrary [65, 77].

However, in some fields of application, procedures which constitute particular cases of the approximation by complete systems of solutions have presented severe restrictions and inconveniences. For the case of acoustics and electromagnetic field computations, a survey of such difficulties was carried out by Bates [78]. In this kind of study the so-called 'Rayleigh hypothesis' restricts drastically the applicability of the method. In view of the previous discussion one may suspect that these difficulties are mainly due to lack of clarity. Indeed, Millar [79, 80] avoided Rayleigh hypothesis altogether by adopting a least-squares approach. Work by other authors has similar implications (see, for example, [81]).

Motivated by this situation and the considerable attention that boundary methods are receiving at present [58, 59] the author started a systematic study of Trefftz's method (more precisely, of the use of complete systems of solutions). The aim of the research was twofold: firstly, to clarify the theoretical foundations required for using complete systems of solutions in a reliable manner; secondly, to expand the versatility of such methods, making them applicable to any problem which is governed by partial differential equations that are linear.

The algebraic theory presented in this book has been found quite suitable for the systematic formulation of the method. The general theory of partial differential equations [82, 83] is fundamental to the developments presented. In connection with Trefftz's method, the following aspects must be covered by the theory:

(a) algorithms for computing the sought solution in the region and on its boundary;
(b) conditions under which they converge;
(c) development of criteria for completeness of a given system of solutions;
(d) general methods for developing complete systems of solutions;
(e) change of boundary conditions;
(f) change of region;
(g) free boundary problems.

Boundary value problems are formulated abstractly as problems with linear constraints. To solve such problems, two algorithms are discussed; one is essentially a least-squares fitting and the other is a generalization of an algorithm which seems to have been originated by a group of Italian mathematicians [84]. The first of these algorithms permits approximation of the solution at any interior point of the region considered; the second is suitable for deriving the boundary values.

The most direct criterion of completeness for a system of solutions of a homogeneous partial differential equation is completeness with respect to the metric of the space in which the operator is defined. For applications to boundary value problems this must be related to the metric of suitable spaces of boundary values. Also, characterization of the completeness of a system of solutions solely by means of boundary values is useful in that it frequently simplifies the analysis, since it avoids introducing a metric which is defined in the whole region. A criterion of completeness possessing these properties is c-completeness or T-completeness (after Trefftz).† Under very general conditions the algorithms proposed converge whenever the system used is T-complete.

The T-completeness criterion supplies considerable flexibility to the procedures for developing complete systems of solutions. Illustrations of this approach are given in Chapter 11. However, no systematic presentation of general methods for developing complete systems of solutions is included.

The property of being T-complete is independent of the specific manner in which the boundary values are imbedded in a Hilbert-space structure, as long as some basic conditions are satisfied; this yields the notion of standard imbedding. However, the function space spanned by the same T-complete system depends on the specific imbedding. This result implies that the same T-complete system can be used, independently of the kind of boundary conditions that are imposed, in most cases of interest. Standard imbeddings are discussed in Chapters 10 and 11.

Questions (d), (f) and (g) are being studied at present. A preliminary discussion of (d) has already appeared [41]. Also some progress has been made on the treatment of free boundary problems by Trefftz's method [42, 44]. Regarding question (f), about conditions under which a T-complete system remains T-complete when the region is changed, there is a good number of contributions on the subject that have been made by several authors [60–65, 68]. However, general criteria applicable to wide classes of partial differential equations are lacking. Some preliminary results,

† The incorporation of this terminology was proposed by Professor O. C. Zienkiewicz, whose suggestion is here gratefully acknowledged.

not yet published, offer some promising indications for formulating general results on this topic in the framework of the theory presented here. But it will be necessary to carry out further research to draw more definite conclusions.

The third type of applications concerns the development of biorthogonal systems of functions to obtain generalized Fourier series. Biorthogonal systems of functions which occur when applying the method of separation of variables to fourth-order equations, such as the biharmonic equation, have received much attention in recent years (see, for example, [85–87]).

In general, the procedure followed by most authors consists in exhibiting a differential equation satisfied by the boundary values of any solution. Then the adjoint of this differential equation is constructed and it is shown that the eigenfunctions of these two systems are biorthogonal. In this manner a formal expansion for any solution of the original differential equation is obtained in which the coefficients are easily computed by means of the biorthogonality relation. Further analysis is required in order to establish the completeness of the system of biorthogonal functions and the convergence of the expansion. Apparently, Smith [88] was the first to deal with these problems successfully. Joseph [85] has exhibited in some recent work the considerable generality of the method, by applying it to a good sample of different problems.

This procedure, however, is not satisfactory in some respects. In particular, the development of the differential equation for the boundary values and its adjoint has an *ad hoc* character which bears little or no relation to the physical situation at hand.

In geophysical studies an independent approach has been followed, also to obtain biorthogonal systems of functions. Indeed, in this field Herrera's [89] orthogonality relations for Rayleigh waves have been known since 1964 [90, 91]. The argument used by Herrera to derive such relations allows complete generality, if suitably formulated, but had remained unnoticed until recently by researchers working in other kinds of applications. Herrera and Spence [37] have explained how the algebraic theory can be used to generalize Herrera's [89] procedure in order to obtain biorthogonal systems of solutions for a wide class of equations. Essentially, the method consists, given a partial differential equation which is linear and homogeneous, in considering the linear space of solutions N_P. Under quite general conditions, the space N_P can be decomposed in two linear subspaces N_P^1 and N_P^2, which are commutative for an antisymmetric bilinear form A_0. Generally, product form solutions yield two families $\{w_1, w_2, \ldots\} \subset N_P^1$ and $\{w_1^*, w_2^*, \ldots\} \subset N_P^2$, which are necessarily biorthogonal with respect to the bilinear form A_0. When the systems of biorthogonal solutions are T-complete, i.e., when, for every $u \in N_P$,

$$\langle A_0 u, w_\alpha \rangle = 0, \quad \forall\, \alpha = 1, 2, \ldots \Rightarrow u \in N_P^1,$$

arbitrary solutions can be developed in a direct manner.

From the perspective of the algebraic theory, the biorthogonal systems are associated with corresponding canonical decompositions of the space of solutions N_P. The bilinear functional A_0, which defines the biorthogonality relation, is clearly related to the partial differential equations considered, and it is derivable from the corresponding operator by integration by parts; no auxiliary adjoint differential system is required. The relation between T-completeness and the notion of Hilbert space bases is well established (Chapter 11). The generality of the procedure is wide, it is not restricted by the order of the partial differential equations, and it is applicable to equations with variable coefficients, as in the case of Herrera's [89] orthogonality relations for Rayleigh waves, which hold for wave-guides with arbitrary transversal heterogeneity. Finally, the biorthogonality property appears as a relation that is satisfied by pairs of solutions in the whole region where they are defined, rather than just at the boundary. This is specially useful when matching solutions in different regions or when modifying the region of definition.

The basic notions and notations are introduced in Chapter 2, and abstract Green's formulas and the general algebraic structure in Chapters 3 and 4. An abstract setting, useful for supplying a general formulation of problems defined in discontinuous fields with prescribed jump conditions—the problem of matching or connecting—is given in Chapter 5. Examples of Green's formulas are given in Chapters 6 and 7.

Part 2 is devoted to boundary methods: more specifically, to Trefftz's method. The scope of Trefftz's method is presented in Chapter 8 and an abstract definition of the space I_P, which characterizes boundary values associated with solutions of homogeneous equations, is introduced in Chapter 9. A general procedure for incorporating the theory in the Hilbert space setting is given in Chapter 10. Generally, every immersion in a Hilbert space can be associated with a Green's formula; thus, one has an immersion of the type discussed in Chapter 10 corresponding to every Green's formula. The T-completeness criterion is studied in Chapter 11, where the relation between this notion, which is based on the boundary values, and completeness for a space of functions defined in the whole region is analyzed. An abstract setting is first introduced and its implications in specific applications are illustrated. The examples given are quite wide, since the most general elliptic equation which is formally symmetric is included. Algorithms for obtaining the solution of boundary value and initial-boundary value problems are given in Chapter 12.

PART 1

Algebraic theory

2 Preliminary notions and notations

Let D be a linear space over the field $\mathfrak{F}$ whose elements will be called scalars [15]. Elements of D will be denoted by $u, v, \ldots$. Write D^* for the linear space of linear functionals defined on D; i.e. D^* is the algebraic dual of D. Hence, any element $\alpha \in D^*$ is a function $\alpha : D \to \mathfrak{F}$ which is linear. Given $v \in D$, the value of the function α at v will be denoted by

$$\alpha(v) = \langle \alpha, v \rangle \in \mathfrak{F}. \tag{2.1}$$

In this work functional-valued operators,

$$P : D \to D^*, \tag{2.2}$$

will be extensively used. Given $u \in D$, the value $P(u) \in D^*$ is itself a linear functional. According to (2.1), given any $v \in D$, $\langle P(u), v \rangle \in \mathfrak{F}$ will be the value of this linear functional at v. When the operator P is itself linear, $\langle P(u), v \rangle$ is linear in u when v is kept fixed. Therefore, as is customary, we write

$$\langle Pu, v \rangle = \langle P(u), v \rangle \in \mathfrak{F} \tag{2.3}$$

for this value. In this work we shall be concerned exclusively with functional valued operators that are linear.

On the other hand, let $D^2 = D \oplus D$ be the space of pairs $[u, v]$ with $u \in D$ and $v \in D$. We may consider functions $\beta : D^2 \to \mathfrak{F}$. The value of such a function on a pair $[u, v] \in D^2$, will be written as $\beta(u, v)$. Such a function is said to be a bilinear functional if it is linear in u when $v \in D$ is kept fixed; conversely, it is linear in v when u is kept fixed [92].

Observe that given any functional valued operator $P : D \to D^*$ which is linear, one can define a bilinear functional $\beta : D^2 \to \mathfrak{F}$ by means of

$$\beta(u, v) = \langle Pu, v \rangle. \tag{2.4}$$

Conversely, given any bilinear functional $\beta : D^2 \to \mathfrak{F}$, one can associate with it an operator $P : D \to D^*$ which is linear. Indeed, given any $u \in D$, let

$$P(u) = \alpha \in D^*, \tag{2.5}$$

where $\alpha \in D^*$ is defined as the linear functional whose value, at any $v \in D$, is

$$\langle \alpha, v \rangle = \beta(u, v). \tag{2.6}$$

Then $P: D \to D^*$ is linear. This establishes a one-to-one correspondence between bilinear functionals and operators $P: D \to D^*$ that are linear. Usually, we will define operators $P: D \to D^*$ by giving their corresponding bilinear functionals.

Given $P: D \to D^*$, take $\beta: D^2 \to \mathfrak{F}$ as the bilinear functional (2.4). Define the operator $P^*: D \to D^*$ as the one associated with the transposed β^* of the bilinear functional β; i.e.,

$$\langle P^*u, v \rangle = \beta^*(u, v) = \beta(v, u) = \langle Pv, u \rangle. \tag{2.7}$$

Throughout this book the notion of 'adjoint' will be used in the sense of 'transposed'. Thus, P^* as defined by (2.7) will be called the adjoint of P. Note that given $P: D \to D^*$, $P^*: D \to D^*$ always exists and is a mapping of the same kind as $P: D \to D^*$. The reason for the simplicity achieved in this manner, is that the notion of adjoint we have just introduced is purely algebraic and can be applied in any linear space D devoid of any further structure, such as a metric or an inner product.

Example 2.1. Take [82, 83] the linear space $D = H^s(\Omega)$, $s \geqslant 2$, where Ω is a bounded region of a finite-dimensional space with boundary $\partial\Omega$ (Fig. 2.1) and define $P: D \to D^*$ by

$$\langle Pu, v \rangle = \int_\Omega v \Delta u \, d\mathbf{x}. \tag{2.8}$$

The adjoint of P is given by

$$\langle P^*u, v \rangle = \int_\Omega u \Delta v \, d\mathbf{x} = \langle Pu, v \rangle + \int_{\partial\Omega} \left\{ u \frac{\partial v}{\partial n} - v \frac{\partial u}{\partial n} \right\} d\mathbf{x}. \tag{2.9}$$

The last equality is obtained by integrating by parts.

Example 2.2. More generally, if $\mathscr{L}$ is a differential operator of order n and $D = H^s(\Omega)$, $s \geqslant n$, one can define $P: D \to D^*$ by

$$\langle Pu, v \rangle = \int_\Omega v \mathscr{L} u \, d\mathbf{x}. \tag{2.10}$$

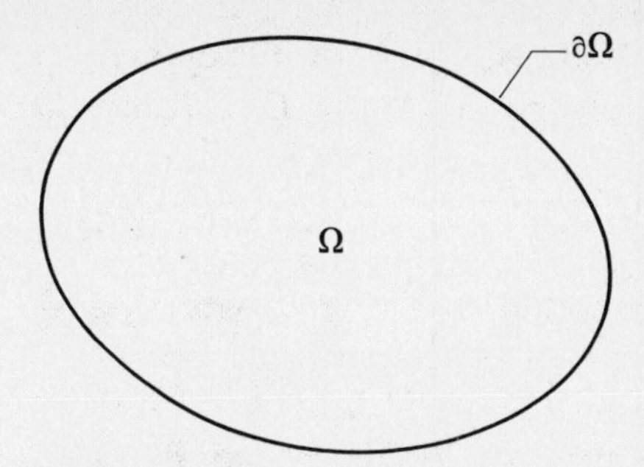

Fig. 2.1 Region of definition of functions belonging to the linear space D

Here it is assumed that the coefficients of $\mathscr{L}$ are infinitely differentiable in Ω.

For any operator $P: D \to D^*$ the null subspace of P will be denoted by N_P. Hence

$$N_P = \{u \in D \mid Pu = 0\}. \tag{2.11}$$

Some relations between the null subspaces of functional-valued operators will be used in the sequel.

Definition 2.1. *One says that the operators $P: D \to D^*$ and $Q: D \to D^*$ can be varied independently, when*

$$D = N_P + N_Q \tag{2.12}$$

The following result clarifies the motivation of this nomenclature.

Lemma 2.1. *Let $P: D \to D^*$ and $Q: D \to D^*$ be linear operators. Then the following assertions are equivalent.*

(a) *P and Q can be varied independently.*

For every $U \in D$ and $V \in D$:

(b) $\exists\, u \in D_\ni \quad Pu = PU \quad \& \quad Qu = QV,$ (2.13)

(c) $\exists\, u \in D_\ni \quad Pu = PU \quad \& \quad Qu = 0,$ (2.14)

(d) $\exists\, u \in D_\ni \quad Pu = 0 \quad \& \quad Qu = QV.$ (2.15)

Proof. Assume that (2.12) holds; given $U \in D$ and $V \in D$, write $U = U_1 + U_2$ and $V = V_1 + V_2$ where U_1, $V_1 \in N_P$ while U_2, $V_2 \in N_Q$. Then $u = V_1 + U_2$ satisfies (2.13). Hence (a) implies (b). It is clear that (b) implies (c) and (d). In view of the symmetric role played by P and Q in Lemma 2.1, to finish its proof it is enough to show that (c) implies (a). Assume (c), then given $v \in D$, take $v_2 \in D$ such that

$$Pv_2 = Pv \quad \& \quad Qv_2 = 0. \tag{2.16}$$

Define $v_1 = v - v_2$. Then, it is seen that $v = v_1 + v_2$, with $v_1 \in N_P$ while $v_2 \in N_Q$. This completes the proof of Lemma 2.1.

Example 2.3. To illustrate Definition 2.1, take P and D as in Example 2.1 and let $B: D \to D^*$ be

$$\langle Bu, v\rangle = -\int_{\partial\Omega} u \frac{\partial v}{\partial n} \, d\mathbf{x}. \tag{2.17}$$

It will be shown that B and B^* can be varied independently. Clearly $B: D \to D^*$ is well-defined when $s \geqslant 2$, because $u \in H^{s-(1/2)}(\partial\Omega)$ while $\partial v/\partial n \in H^{s-(3/2)}(\partial\Omega)$ [82]. Therefore the same is true of $B^*: D \to D^*$. Moreover, the ranges of the traces u and $\partial u/\partial n$ on $\partial\Omega$ are $H^{s-(1/2)}(\partial\Omega)$ and $H^{s-(3/2)}(\partial\Omega)$, respectively, when u ranges over $H^s(\Omega)$. Thus

$$u \in N_B \Leftrightarrow \int_{\partial\Omega} u \frac{\partial v}{\partial n} \, d\mathbf{x} = 0 \quad \forall \frac{\partial v}{\partial n} \in H^{s-(3/2)}(\partial\Omega). \tag{2.18}$$

Therefore

$$N_B = \{u \in D \mid u = 0 \quad \text{on} \quad \partial\Omega\}, \tag{2.19a}$$

since $H^{s-(3/2)}(\partial\Omega) \subset H^{\circ}(\partial\Omega)$ is dense in $H^{\circ}(\partial\Omega)$. In a similar fashion it can be seen that

$$N_{B^*} = \left\{u \in D \,\middle|\, \frac{\partial u}{\partial n} = 0 \quad \text{on} \quad \partial\Omega\right\}. \tag{2.19b}$$

Now [82], the range of the pairs of traces $[u, (\partial u/\partial n)]$ on $\partial\Omega$, when u ranges over $H^s(\Omega)$, is $H^{s-(1/2)}(\partial\Omega) \oplus H^{s-(3/2)}(\partial\Omega)$. This implies that given any $u \in H^s(\Omega) = D$, there exist functions $u_1 \in D$ and $u_2 \in D$ such that $u = u_1 + u_2$, but

$$u_1 = 0 \quad \text{and} \quad \frac{\partial u_2}{\partial n} = 0, \quad \text{on} \quad \partial\Omega, \tag{2.20}$$

Thus, $u_1 \in N_B$ and $u_2 \in N_{B^*}$.

Example 2.4. As an illustration of Lemma 2.1, take $D = H^s(\Omega)$,

$s \geqslant 2$, and let $P: D \to D^*$ and $B: D \to D^*$ be given by equations (2.8) and (2.17), respectively. Then, it is easy to verify that condition (d) of Lemma 2.1 is satisfied with Q replaced by B. Indeed, it is straightforward to see that

$$N_P = \{u \in D \mid \Delta u = 0 \quad \text{in} \quad \Omega\}. \tag{2.21}$$

This equation together with (2.19a) implies that condition (d) is equivalent to the existence of a solution for the Dirichlet problem.

$$\Delta u = 0 \text{ on } \Omega, \quad \text{with} \quad \gamma_0 u = \gamma_0 V \in H^{s-(1/2)}(\partial\Omega). \tag{2.22}$$

Here, as is standard [83], γ_0 has been used for the trace of u on $\partial\Omega$. Well-known results on the existence of a solution for elliptic equation [82] grant the existence of a solution for this problem. Hence, condition (d) is satisfied; i.e., B and P can be varied independently.

Assume $P: D \to D^*$, $Q: D \to D^*$ and $R: D \to D^*$ are such that

$$P = Q + R. \tag{2.23}$$

Then it is straightforward to see that

$$N_P \supset N_Q \cap N_R \tag{2.24}$$

A stronger result holds when Q^* and R^* can be varied independently.

Lemma 2.2. *Assume P, Q and R are such that $P = Q + R$, while Q^* and R^* can be varied independently. Then*

$$N_P = N_Q \cap N_R. \tag{2.25}$$

Proof. In view of (2.24), it is only necessary to prove that when Q^* and R^* can be varied independently, $N_P \subset N_Q \cap N_R$.

Assume $PU = 0$ and $QU \neq 0$. Then, $\exists\ V \in D_\ni \langle QU, V\rangle \neq 0$. For such V, take $v \in D_\ni Q^*v = Q^*V$ while $R^*v = 0$. Recall that

$$\begin{aligned} 0 &\neq \langle QU, V\rangle = \langle Q^*V, U\rangle = \langle Q^*v, U\rangle + \langle R^*v, U\rangle \\ &= \langle QU, v\rangle + \langle RU, v\rangle = \langle PU, v\rangle = 0. \end{aligned} \tag{2.26}$$

Hence, $U \in N_P \Rightarrow U \in N_Q$; i.e., $N_P \subset N_Q$. A similar argument, replacing Q by R, yields $N_P \subset N_R$. Hence $N_P \subset N_Q \cap N_R$ and the lemma is established.

Example 2.5. Let $D = H^s(\Omega)$, $s \geq 2$, and define

$$\langle Au, v\rangle = -\int_{\partial\Omega} \left\{ u\frac{\partial v}{\partial n} - v\frac{\partial u}{\partial n} \right\} d\mathbf{x} = \langle (B - B^*)u, v\rangle. \tag{2.27}$$

Then

$$A = B - B^*, \tag{2.28}$$

where B is given by (2.17). Example 2.3 shows that B^* and $(-B^*)^* = -B$ can be varied independently. Hence

$$N_A = N_B \cap N_{B^*} = \{u \in D \mid u = \partial u/\partial n = 0 \quad \text{on} \quad \partial\Omega\}. \tag{2.29}$$

3 Formal adjoints and Green's formulas

The notions of formal adjoints and Green's formulas are usually defined for differential operators. It is possible, however, to introduce such concepts for general functional valued operators of the type $P:D\to D^*$. The corresponding theory is developed in this chapter.

Definition 3.1. *$B:D\to D^*$ is a boundary operator for $P:D\to D^*$ when*

$$\langle Pu, v\rangle = 0 \quad \forall\, v\in N_B \Rightarrow Pu=0. \tag{3.1}$$

Example 3.1. Let the operators $B:D\to D^*$ ($D=H^s(\Omega)$, $s\geq 2$) and $P:D\to D^*$ be given by (2.17) and (2.8), respectively. Then it is easy to see that B is a boundary operator for P in the sense of Definition 3.1. Indeed, in this case, condition (3.1) becomes

$$\int_\Omega v\Delta u \,\mathrm{d}\mathbf{x} = 0 \quad \forall\, v\in N_B \Rightarrow Pu=0. \tag{3.2}$$

Now, $N_B\supset \mathring{H}^s(\Omega)$ by virtue of (2.19a). But $\mathring{H}^s(\Omega)$ is dense in $H^\circ(\Omega)$ and the premise in (3.2) implies that $\Delta u\in H^{s-2}(\Omega)\subset H^\circ(\Omega)$ is orthogonal (in the $\mathcal{L}^2=H^\circ$ sense), to every element $v\in N_B$. Hence $\Delta u\equiv 0$ and $u\in N_P$ by (2.8).

Definition 3.2. *Given $P:D\to D^*$ and $Q:D\to D^*$, define $S=P-Q^*$. Then P and Q are formal adjoints if S is a boundary operator for Q while S^* is a boundary operator for P. Notice that this is a symmetric relation between P and Q as can be easily verified.*

Example 3.2. Let $D=H^s(0,T)$, $s\geq 1$, where $(0,T)$ refers to the open interval of the real line. Define $P:D\to D^*$ by

$$\langle Pu, v\rangle = \int_0^T v\frac{\mathrm{d}u}{\mathrm{d}t}\,\mathrm{d}t. \tag{3.3}$$

Take

$$Q = -P. \tag{3.4}$$

Then

$$\langle Su, v\rangle = \int_0^T \frac{\mathrm{d}uv}{\mathrm{d}t}\,\mathrm{d}t = u(T)v(T) - u(0)v(0). \tag{3.5}$$

Observe that $S^* = S$. Also

$$N_S = N_{S^*} = \{u \in D \mid u(0) = u(T) = 0\}. \tag{3.6}$$

Using (3.6) it is easy to see that S^* is a boundary operator for P, while S is a boundary operator for $-Q$. Hence P and $-P$ are formal adjoints.

Definition 3.3. *An operator $P: D \to D^*$ is said to be formally symmetric when P is a formal adjoint of itself.*

Theorem 3.1. *Given $P: D \to D^*$, define*

$$A = P - P^*. \tag{3.7}$$

Then P is formally symmetric, if and only if

$$\langle Pu, v\rangle = 0 \quad \forall\, v \in N_A \Rightarrow Pu = 0. \tag{3.8}$$

Proof. Set $Q = P$ in Definition 3.2. Then $S = P - P^* = A$ and P is formally symmetric if and only if S and $S^* = -S$ are boundary operators for P. Clearly this is equivalent to (3.8).

Theorem 3.2. *Given $P: D \to D^*$ and $Q: D \to D^*$, define $\hat{P}: \hat{D} \to \hat{D}^*$ by*

$$\langle \hat{P}\hat{u}, \hat{v}\rangle = \langle Pu_1, v_2\rangle + \langle Qu_2, v_1\rangle, \tag{3.9}$$

where $\hat{D} = D \oplus D$ and elements $\hat{u} \in \hat{D}$ are denoted by $\hat{u} = [u_1, u_2]$, $u_1 \in D$, $u_2 \in D$. Then P and Q are formal adjoints, if and only if $\hat{P}$ is formally symmetric.

Proof. The following lemma can be used to prove this theorem.

Lemma 3.1. *Let*

$$\hat{A} = \hat{P} - \hat{P}^*. \tag{3.10}$$

Then

$$\hat{N}_A = N_S \oplus N_{S^*}, \tag{3.11}$$

where $S = P - Q^*$ *and* $\hat{N}_A$ *is the null subspace of* $\hat{A}$.

Proof. For every $\hat{u} \in \hat{D}$ and $\hat{v} \in \hat{D}$, one has

$$\langle \hat{A}\hat{u}, \hat{v} \rangle = \langle Su_1, v_2 \rangle - \langle S^* u_2, v_1 \rangle. \tag{3.12}$$

Thus $\hat{u} \in \hat{N}_{\hat{A}} \Leftrightarrow u_1 \in N_S$ while $u_2 \in N_{S^*}$. This completes the proof of the lemma.

According to Theorem 3.1 a necessary and sufficient condition for $\hat{P}$ to be formally symmetric is that

$$\langle \hat{P}\hat{u}, \hat{v} \rangle = 0 \quad \forall\, \hat{v} \in \hat{N}_{\hat{A}} \Rightarrow \hat{P}\hat{u} = 0. \tag{3.13}$$

In view of Lemma 3.1 this is equivalent to

$$\langle Pu_1, v_2 \rangle = 0 \quad \forall\, v_2 \in N_{S^*} \Rightarrow Pu_1 = 0 \tag{3.14}$$

and simultaneously

$$\langle Qu_2, v_1 \rangle = 0 \quad \forall\, v_1 \in N_S \Rightarrow Qu_2 = 0. \tag{3.15}$$

Clearly, (3.14) means that S^* is a boundary operator for P, while (3.15) means that S is a boundary operator for Q. Application of Definition 3.2 yields the desired result.

The following examples illustrate the fact that differential operators which are formally symmetric in the classical sense can always be associated with functional-valued operators which are formally symmetric in the sense introduced here.

Example 3.3. Taking P and D as in Example 2.1, it is seen that

$$\langle Au, v \rangle = \int_{\partial\Omega} \left\{ v \frac{\partial u}{\partial n} - u \frac{\partial v}{\partial n} \right\} \mathrm{d}\mathbf{x}. \tag{3.16}$$

Hence

$$N_A = \left\{ u \in D \mid u = \frac{\partial u}{\partial n} = 0, \quad \text{on} \quad \partial\Omega \right\}. \tag{3.17}$$

The argument used to prove (3.2) can now be applied to show (3.8), because $N_A \supset H_0^s(\Omega)$. Hence, $P: D \to D^*$ is formally symmetric.

Example 3.4. The proof is given here of the assertion that any differential operator that is self-adjoint in the classical sense can be associated with an operator $P:D \to D^*$ which is formally symmetric.

If†

$$\mathscr{L}u = \sum_{|p|,|q|\leq m} (-1)^{|p|} D^p(a_{qp}(x) D^q u), \tag{3.18}$$

where $a_{qp} \in \mathfrak{D}(\bar{\Omega})$ and $a_{qp} = a_{pq}$. Then, if $D = H^s(\Omega)$, $s \geqslant 2m$, it can be seen that the operator $P:D \to D^*$, given by

$$\langle Pu, v\rangle = \int_\Omega v\mathscr{L}u \, \mathrm{d}\mathbf{x}, \tag{3.19}$$

is formally symmetric. Indeed, if $u \in H_0^s(\Omega)$, $s \geqslant 2m$, by integration by parts, it is seen that

$$\langle Au, v\rangle = \langle Pu, v\rangle - \langle Pv, u\rangle = 0 \quad \forall\, v \in D, \tag{3.20}$$

because the boundary terms coming from the integration by parts [82, p. 115] vanish when $u \in H_0^s(\Omega)$. This shows that $N_A \supset H_0^s(\Omega)$ and the result is straightforward.

Example 3.5. It is interesting to observe that it is possible to associate formally symmetric operators $P:D \to D^*$ to differential operators which are not formally symmetric in the classical sense. Take $D = H^s[(0, T)]$, $s \geqslant 1$ and define $P:D \to D^*$ by

$$\langle Pu, v\rangle = v * \frac{\mathrm{d}u}{\mathrm{d}t} = \int_0^T v(T-t)\frac{\mathrm{d}u}{\mathrm{d}t}(t)\,\mathrm{d}t. \tag{3.21}$$

Then

$$\langle Au, v\rangle = \langle (P-P^*)u, v\rangle = v(0)u(T) - v(T)u(0). \tag{3.22}$$

Hence

$$N_A = \{u \in D \mid u(0) = u(T) = 0\}. \tag{3.23}$$

Again, $N_A \supset H_0^s[(0, T)]$ and $P:D \to D^*$ is formally symmetric, because implication (3.8) is satisfied.

The introduction of formally symmetric operators of the type $P:D \to D^*$, associated with differential operators which are not formally symmetric in the classical sense, is of interest on several

† The notation used here is that of Lions and Magenes [82, pp. 1, 109 and 114].

counts. One of them is in connection with the development of variational principles; as a matter of fact, this is the basis of a class of principles known in the literature as Gurtin's variational principles [14, 47, 48, 50]. The following example presents an operator which is suitable for the discussion of the heat equation.

Example 3.6. Let $D=H^p(0,T;H^s(\Omega))$, with $p\geq 1$ and $s\geq 2$. Define

$$\langle Pu, v\rangle = \int_{\partial\Omega} v * \left\{\frac{\partial u}{\partial t} - \Delta u\right\} d\mathbf{x}, \tag{3.24}$$

where the operation $u * v$ is defined in (3.21). It is easy to verify that $P:D\to D^*$, as given by (3.24), is formally symmetric.

In the standard approach the wave equation is associated with the operator

$$\langle Pu, v\rangle = \int_0^T \int_\Omega v\left\{\frac{\partial^2 u}{\partial t^2} - \Delta u\right\} d\mathbf{x}\, dt. \tag{3.25}$$

Applying the criterion (3.8), this operator can be shown to be formally symmetric. Using it, one can develop Green's formulas which are suitable for application to boundary value problems. However they cannot be used to treat initial value problems. For this latter purpose it is more convenient to introduce, instead,

$$\langle Pu, v\rangle = \int_{\partial\Omega} v * \left\{\frac{\partial^2 u}{\partial t^2} - \Delta u\right\} d\mathbf{x}. \tag{3.26}$$

This operator can also be seen to be formally symmetric in the sense of Definitions 3.2 and 3.3.

The notion of Green's formula for formally symmetric operators is given next. The introduction of a more general notion of Green's formula, applicable to operators which may not be formally symmetric, will not be discussed in this monograph.

Definition 3.4. *Assume that*

(i) $P:D\to D^*$ *is formally symmetric;*
(ii) B *and* B^* *are boundary operators for* P;
(iii) B *and* B^* *can be varied independently.*
(iv) *The equation*

$$A = P - P^* = B - B^* \tag{3.27}$$

is satisfied. Then equation (3.27) *is said to be a Green's formula.*

It is interesting to introduce an additional notion which is closely related.

Definition 3.5. *An operator $B:D\to D^*$ is said to decompose* A, *when B and B^* can be varied independently, and they satisfy equation* (3.27).

Clearly, when (3.27) is a Green's formula, B decomposes A. When $P:D\to D^*$ is formally symmetric the converse is also true.

Theorem 3.3 *Let $P:D\to D^*$ be formally symmetric. Assume that $B:D\to D^*$ decomposes* A. *Then equation* (3.27) *is a Green's formula.*

Proof. It is only required to show that B and B^* are boundary operators. This is immediate because

$$N_A = N_B \cap N_{B^*} \tag{3.28}$$

by virtue of Lemma 2.2. Hence, $N_B \supset N_A$ and $N_{B^*} \supset N_A$. In view of Definition 3.1, Theorem 3.1 implies now the desired result.

Example 3.7. The operators P and B considered in Example 2.4 are such that B decomposes A. Indeed

$$\begin{aligned}\langle (P-P^*)u, v\rangle &= -\int_\Omega \{(u\Delta v)-(v\Delta u)\}\,\mathrm{d}\mathbf{x}\\ &= -\int_{\partial\Omega}\left\{u\frac{\partial v}{\partial n}-v\frac{\partial u}{\partial n}\right\}\mathrm{d}\mathbf{x}\\ &= \langle (B-B^*)u, v\rangle.\end{aligned} \tag{3.29}$$

Hence

$$A = P - P^* = B - B^*. \tag{3.30}$$

In Example 2.3, it was shown that B and B^* can be varied independently.

Example 3.8. For the operator introduced in Example 2.1, it is easy to derive general Green's formulas which are defined pointwise. It can be seen that, in this case, the fact that B and B^* can be varied independently is granted if

$$\langle Bu, v\rangle = \int_{\partial\Omega}\left(b_1 v + b_2\frac{\partial v}{\partial n}\right)\left(a_1 u + a_2\frac{\partial u}{\partial n}\right)\mathrm{d}\mathbf{x}, \tag{3.31}$$

where a_1, a_2, b_1 and b_2 can be taken as functions of position on $\partial\Omega$. Restrictions associated with the degree of smoothness that these functions must satisfy will be discussed later. Equation (3.27) holds, if and only if

$$a_1 b_2 - a_2 b_1 = -1. \tag{3.32}$$

This grants that B and B^* can be varied independently. More general Green's formulas which are not defined pointwise will be discussed later.

4 Abstract characterization of Green's formulas. Regular subspaces and canonical decompositions

At the end of Chapter 3 it was shown that, for formally symmetric operators, Green's formulas (Definition 3.4) are characterized by operators B that decompose A (Definition 3.5); more precisely, that (3.27) is a Green's formula if and only if B decomposes A. Here, it will be shown that operators B that decompose A can be defined by means of the null subspaces N_B and N_{B^*}. This yields an abstract characterization of Green's formulas which supplies considerable flexibility to the algebraic theory presented in this book.

Some properties of the operator $A = P - P^*$ introduced in Theorem 3.1 play an important role in the developments that follow. In particular, $A : D \to D^*$ is skew-symmetric (or antisymmetric); i.e.,

$$\langle Au, v\rangle = -\langle Av, u\rangle \quad \forall\, u \in D \quad \& \quad v \in D. \tag{4.1}$$

Generally, for the results to hold it is not essential to have an operator $P : D \to D^*$ defined such that $A = P - P^*$, as long as (4.1) is satisfied. The structure presented in this chapter is closely related to symplectic geometry [45, 46]; however, in this latter theory the assumption that $N_A = \{0\}$, which will not be incorporated here, is usually adopted.

If $A = P - P^*$ and P is formally symmetric, A is a boundary operator for P. In such case, the operator A can be used to introduce a classification of boundary values. Indeed, given $u \in D$ one can characterize the boundary values of u by means of the functional $Au \in D^*$. In the definition that follows this notion is introduced in a slightly more general context.

Definition 4.1. *Let $A : D \to D^*$ be skew-symmetric. The relevant boundary values of $u \in D$ and $v \in D$ are said to be equal, if and only if $Au = Av$.*

Since A is linear, this definition is tantamount to saying that the relevant boundary values of u and v are equal, if and only if

$u - v \in N_A$. Therefore, this classification leads to consideration of the quotient space $\mathfrak{D} = D/N_A$, i.e. the space of cosets $u + N_A \subset D$, with $u \in D$ [93].

Example 4.1. When $A: D \to D^*$ is given by (3.16) the null subspace N_A is (equation 3.17)

$$N_A = \left\{ u \in D \mid u = \frac{\partial u}{\partial n} = 0, \quad \text{on} \quad \partial\Omega \right\}. \tag{4.2}$$

The classification of boundary values is characterized by the pair of functions u and $\partial u/\partial n$, defined on $\partial\Omega$.

It is usual [46] to say that a linear subspace I is isotropic when

$$\langle Au, v \rangle = 0 \quad \forall\, u \in I \quad \& \quad v \in I. \tag{4.3}$$

A further condition will be useful in further developments.

Definition 4.2. *An isotropic subspace $I \subset D$ is said to be regular when*

$$I \supset N_A. \tag{4.4}$$

Notice that a subspace that satisfies (4.4) is characterized by relevant boundary values only; alternatively, one can say that any regular subspace I is necessarily a coset, or a union of several cosets, of the quotient space D/N_A. Also, if two functions $u \in D$ and $v \in D$ have the same relevant boundary values and I is regular, then one and only one of the following mutually exclusive possibilities holds:

$$u \text{ and } v \text{ belong to } I \tag{4.5a}$$

or

$$u \text{ and } v \text{ do not belong to } I. \tag{4.5b}$$

On the other hand, isotropic subspaces will also be called commutative. This nomenclature is appropriate because, when (4.3) holds, the restriction to I of the bilinear functional associated with P is commutative.

To summarize our comments, Definition 4.2 states that a regular subspace is a commutative subspace which is characterized by relevant boundary values only.

Example 4.2. When A is given by (3.16),

$$I=\{u\in D \mid u=0 \quad \text{on} \quad \partial\Omega\} \tag{4.6}$$

is a regular subspace.

Definition 4.3. *A linear subspace $I\subset D$ is said to be completely regular for P, when for every $u\in D$, one has*

$$\langle Au, v\rangle=0 \quad \forall\, v\in I \Leftrightarrow u\in I. \tag{4.7}$$

Remark 4.1. This definition is related to the notion of Lagrangean subspace used in symplectic geometry [46]. Indeed, for the case $N_A=\{0\}$ they can be shown to be equivalent.

An alternative manner of defining the concept of completely regular subspace is as a commutative subspace that is largest. The precise meaning of this statement is given next.

Lemma 4.1. *A linear subspace I, which is commutative, is completely regular, if and only if, for every commutative subspace I', one has*

$$I'\supset I \Rightarrow I'=I. \tag{4.8}$$

Proof. Assume I is completely regular. Let $I'\supset I$ be commutative. Then any $u\in I'$ satisfies the premise in (4.7) which implies $u\in I$. Hence $I\supset I'$ and $I=I'$. Notice that the implication in the left-hand sense contained in (4.7) is equivalent to (4.3), i.e., it simply means that I is commutative. Therefore, to prove the remaining part of Lemma 4.1, it is only required to show that the implication in the right-hand sense in (4.7) is satisfied by any commutative subspace I for which (4.8) holds. Assume $u\notin I$ fulfils the premise in (4.7), then $I'=I+\{u\}$ is commutative and contains I properly; this contradicts (4.8). Hence $u\in I$.

Example 4.3. The subspace $I\subset D=H^s(\Omega)$, $s\geq 2$, given in Example 4.2, is completely regular for A, as given by (3.16). This can be verified easily, because the range of $\partial u/\partial n$ on $\partial\Omega$ is $H^{s-(3/2)}(\partial\Omega)$ which is dense in $H^0(\partial\Omega)$.

Lemma 4.2. *$I\subset D$ is completely regular, if and only if I is a commutative subspace and*

$$\langle Au, v\rangle=0 \quad \forall\, v\in I \Rightarrow u\in I. \tag{4.9}$$

Proof. The equivalence statement (4.7) is the conjunction of (4.3) and (4.9); thus, any regular subspace satisfying (4.9) is completely regular. To prove the converse, it is only necessary to show that when $I \subset D$ is completely regular, (4.4) holds. But this is immediate because, if $u \in N_A$ and I is completely regular, then $\langle Au, v\rangle = 0$ for every $v \in D \supset I$. Hence, $u \in I$ by virtue of (4.9).

In the discussion of boundary methods the following definition and lemma will be useful.

Definition 4.4. *A subset $\mathfrak{B} \subset I$ is said to be T-complete or c-complete (complete in connectivity) for the subspace I, when*

$$\langle Au, w\rangle = 0 \quad \forall\, w \in \mathfrak{B} \Rightarrow u \in I. \tag{4.10}$$

When, in addition, for every finite subset $\{w_1, w_2, \ldots, w_n\} \subset \mathfrak{B}$, the functionals $\{Aw_1, Aw_2, \ldots, Aw_n\}$ are linearly independent, $\mathfrak{B}$ is said to be a connectivity basis.

Lemma 4.3. *Let $I \subset D$ be a commutative subspace. Then I is completely regular if and only if it possesses a T-complete subset.*

Proof. When I is completely regular, I itself is a T-complete subset. Conversely, when there exists $\mathfrak{B} \subset I$ which is T-complete, (4.9) is necessarily satisfied. Using this fact, it is clear that I is completely regular by virtue of Lemma 4.2.

Definition 4.5. *An ordered pair $\{I_1, I_2\}$ of regular subspaces such that*

$$D = I_1 + I_2 \tag{4.11}$$

is said to be a canonical decomposition of D with respect to A (or P).

Remark 4.2. When $N_A = \{0\}$, the notion of canonical decomposition reduces to that of Lagrangean decomposition of symplectic geometry.

Lemma 4.4. *Assume $B : D \to D^*$ decomposes A. Define*

$$I_1 = N_B, \qquad I_2 = N_{B^*}. \tag{4.12}$$

Then the pair $\{I_1, I_2\}$ is a canonical decomposition of D.

Proof. The fact that B and B^* can be varied independently implies (4.11). Assume $u \in I_1$ and $v \in I_1$, then

$$\langle Au, v\rangle = \langle Bu, v\rangle - \langle Bv, u\rangle = 0. \tag{4.13}$$

Also, $I_1 = N_B \supset N_A$ by virtue of (3.28). This shows that I_1 is regular. In a similar fashion it can be shown that I_2 is regular.

Example 4.4. Taking the operators $P: D \to D^*$ and $B: D \to D^*$ as in Examples 2.1 and 2.3, it was shown that B decomposes A (Example 3.7). Therefore the pair of subspaces

$$I_1 = N_B = \{u \in D \mid u = 0, \quad \text{on} \quad \partial\Omega\} \tag{4.14a}$$

and

$$I_2 = N_{B^*} = \{u \in D \mid \partial u/\partial n = 0, \quad \text{on} \quad \partial\Omega\} \tag{4.14b}$$

constitutes a canonical decomposition for P.

The properties involved in Definition 4.5 of canonical decomposition imply stronger properties.

Theorem 4.1. *A pair of subspaces $\{I_1, I_2\}$ is a canonical decomposition of D if and only if I_1 and I_2 are completely regular:*

$$D = I_1 + I_2 \quad \textit{and} \quad N_A = I_1 \cap I_2. \tag{4.15}$$

Proof. In view of Definition 4.5, it is clear that any pair of completely regular subspaces $\{I_1, I_2\}$ that satisfies (4.15) is a canonical decomposition. Thus, only the converse statement needs to be proved. The following lemma will be useful for this purpose.

Lemma 4.5. *Assume that $\{I_1, I_2\}$ is a canonical decomposition of D. Then*

$$u_1 \in I_1 \quad \textit{and} \quad \langle Au_1, v_2\rangle = 0 \quad \forall\, v_2 \in I_2 \Rightarrow u_1 \in N_A. \tag{4.16}$$

Similarly,

$$u_2 \in I_2 \quad \textit{and} \quad \langle Au_2, v_1\rangle = 0 \quad \forall\, v_1 \in I_1 \Rightarrow u_2 \in N_A. \tag{4.17}$$

Proof. To prove (4.16), assume $u_1 \in I_1$ is such that

$$\langle Au_1, v_2\rangle = 0 \quad \forall\, v_2 \in I_2. \tag{4.18}$$

Then, given any $W \in D$, let $W_1 \in I_1$ and $W_2 \in I_2$ be such that $W = W_1 + W_2$. Clearly,

$$\langle Au_1, W\rangle = \langle Au_1, W_1\rangle + \langle Au_1, W_2\rangle = 0. \tag{4.19}$$

This shows $u_1 \in N_A$. That (4.17) also holds is clear, by virtue of the symmetric roles played by I_1 and I_2.

An immediate corollary of Lemma 4.5 is that when $\{I_1, I_2\}$ is a canonical decomposition

$$I_1 \cap I_2 \subset N_A. \tag{4.20}$$

In addition, the converse of (4.20) necessarily holds since $I_1 \supset N_A$ and $I_2 \supset N_A$. Hence

$$I_1 \cap I_2 = N_A. \tag{4.21}$$

Thus, in order to complete the proof of Theorem 4.1, it remains to prove that I_1 and I_2 are completely regular. To this end, given any $u \in D$, write $u = u_1 + u_2$ with $u_1 \in I_1$ and $u_2 \in I_2$. Then, if for every $v_1 \in I_1$

$$\langle Au, v_1 \rangle = \langle Au_2, v_1 \rangle = 0, \tag{4.22}$$

one has $u_2 \in N_A \subset I_1$ by (4.17). Hence $u = u_1 + u_2 \in I_1$. In view of Lemma 4.2, this shows that I_1 is completely regular and a similar argument yields the corresponding result for I_2.

Theorem 4.2. *With every operator $B: D \to D^*$ that decomposes A, associate a canonical decomposition $\{I_1, I_2\}$ by means of (4.12). Then, such correspondence between operators that decompose A and canonical decompositions is one-to-one and covers the set of canonical decompositions of D. Under this mapping, any canonical decomposition $\{I_1, I_2\}$ is the image of a unique operator $B: D \to D^*$ given by*

$$\langle Bu, v \rangle = \langle Au_2, v_1 \rangle, \tag{4.23}$$

where u_2 and v_1 are the components of u and v on I_2 and I_1, respectively.

Proof. We start with an observation. Let $\{I_1, I_2\}$ be a canonical decomposition, then given any $u \in D$ and $v \in D$, $\langle Au_2, v_1 \rangle$ is uniquely defined when $u_2 \in I_2$ and $v_1 \in I_1$ are taken as in (4.23). This follows from Theorem 4.1, since u_2 and v_1 are uniquely defined except by elements of N_A. Clearly, such indeterminacy is irrelevant when evaluating

$$\langle Au_2, v_1 \rangle = -\langle Av_1, u_2 \rangle.$$

Assume $\{I_1, I_2\}$ is a canonical decomposition. If, in addition, there exists $B: D \to D^*$ such that $A = B - B^*$ and satisfying (4.12), then

$$\langle Bu, v\rangle = \langle Bu_2, v\rangle = \langle B^*v, u_2\rangle = \langle B^*v_1, u_2\rangle$$
$$= \langle (B^* - B)v_1, u_2\rangle = -\langle Av_1, u_2\rangle. \tag{4.24}$$

This shows that if such B exists, it is unique and it is given by (4.23). Thus, the correspondence is one-to-one. Once this has been shown, it only remains to prove that when a canonical decomposition $\{I_1, I_2\}$ is given, equation (4.23) defines an operator $B: D \to D^*$ which decomposes A and satisfies (4.12). Recall that B is well defined. Also, equation (3.27) is satisfied because

$$\langle Au, v\rangle = \langle Au_2, v_1\rangle - \langle Av_2, u_1\rangle. \tag{4.25}$$

Clearly, $I_1 \subset N_B$ by virtue of (4.23) and the fact that $u_2 \in N_A$ whenever $u \in I_1$. Assume $u \in N_B$ (i.e., $Bu = 0$), then

$$\langle Au, v_1\rangle = \langle Au_2, v_1\rangle = \langle Bu, v_1\rangle = 0 \quad \forall\, v_1 \in I_1. \tag{4.26}$$

This shows $u \in I_1$, since I_1 is completely regular. Hence $I_1 \supset N_B$ and the first of equations (4.12) holds. A dual argument yields the second one. Once this has been shown, it is clear that

$$D = I_1 + I_2 = N_B + N_{B^*}.$$

Hence B decomposes A and the proof of Theorem 4.2 is complete.

Example 4.5. Let us re-examine Example 3.8 under the light of the results just obtained. Let the linear space D be $\Re^2$ (the two-dimensional Euclidean space). Define $A: D \to D^*$ by

$$\langle A\mathbf{u}, \mathbf{v}\rangle = v_1 u_2 - u_1 v_2, \tag{4.27}$$

where the notation $\mathbf{u} = [u_1, u_2]$ has been adopted for elements of D. Notice that the expression

$$-v\frac{\partial u}{\partial n} + u\frac{\partial v}{\partial n} = v_1 u_2 - u_1 v_2, \tag{4.28}$$

occurring in (3.16), is obtained if the pair $[u_1, u_2]$ is identified with the pair $[u, -(\partial u/\partial n)]$. If the transformation

$$\mathscr{A}\mathbf{u} = [u_2, -u_1] \tag{4.29}$$

is defined, then

$$\langle A\mathbf{u}, \mathbf{v}\rangle = \mathbf{v} \cdot \mathscr{A}\mathbf{u}. \tag{4.30}$$

Clearly, $\mathscr{A}$ is a 90° rotation, clockwise. Also $N_A = \{0\}$ and a subspace I is regular, if and only if

$$I \perp \mathscr{A}I. \tag{4.31}$$

This implies that I is regular if and only if the dimension of I is less or equal to 1. Let I_1 and I_2 be two regular subspaces of dimension 1, then $I_1 + I_2 = D$ if and only if I_1 and I_2 are not parallel. Also, any regular subspace of dimension 1 is completely regular.

Therefore, any canonical decomposition can be characterized by a pair of non-parallel vectors $\{\mathbf{a}, \mathbf{b}\}$, such that

$$I_1 = \{\mathbf{u} \in D \mid \mathbf{a} \cdot \mathbf{u} = 0\}, \tag{4.32a}$$

while

$$I_2 = \{\mathbf{v} \in D \mid \mathbf{b} \cdot \mathbf{v} = 0\}. \tag{4.32b}$$

Observe that the conditions defining I_1 and I_2 are equivalent to

$$\mathscr{A}\mathbf{a} \cdot \mathscr{A}\mathbf{u} = \mathscr{A}\mathbf{b} \cdot \mathscr{A}\mathbf{v} = 0. \tag{4.33}$$

In order to apply formula (4.23), for every $\mathbf{u} \in D$, adopt the notation $\mathbf{u} = \mathbf{u}_1 + \mathbf{u}_2$, with $\mathbf{u}_1 \in I_1$ and $\mathbf{u}_2 \in I_2$. The operator $B : D \to D^*$ is, by definition, a bilinear functional; thus, application of formula (4.23) yields

$$\sum_{i,j \leq 2} C_{ij} u_i v_j = \langle B\mathbf{u}, \mathbf{v} \rangle = \mathbf{v}_1 \cdot \mathscr{A}\mathbf{u}_2. \tag{4.34}$$

Also

$$\sum_{i,j \leq 2} C_{ij} u_i v_j = 0 \quad \forall\, \mathbf{u} \in I_1 \quad \text{and} \quad \forall\, \mathbf{v} \in I_2. \tag{4.35}$$

Equation (4.35) implies

$$C_{ij} = \lambda a_i b_j. \tag{4.36}$$

Here, λ is a scalar. In view of (4.35) and (4.36), equation (4.34) can be written as

$$\lambda(\mathbf{a} \cdot \mathbf{u})(\mathbf{b} \cdot \mathbf{v}) = \lambda(\mathbf{a} \cdot \mathbf{u}_2)(\mathbf{b} \cdot \mathbf{v}_1) = \mathbf{v}_1 \cdot \mathscr{A}\mathbf{u}_2. \tag{4.37}$$

Observe, finally, that every $\mathbf{v}_1 \in I_1$ is parallel to $\mathscr{A}\mathbf{a}$, while every $\mathbf{u}_2$ is parallel to $\mathscr{A}\mathbf{b}$. This shows that (4.37) is satisfied for every $\mathbf{v}_1 \in I_1$ and every $\mathbf{u}_2 \in I_2$, if and only if

$$\lambda = \frac{-1}{\mathbf{a} \cdot \mathscr{A}\mathbf{b}}. \tag{4.38}$$

Hence

$$\langle B\mathbf{u}, \mathbf{v}\rangle = \lambda\left(a_1 u + a_2 \frac{\partial u}{\partial n}\right)\left(b_1 v + b_2 \frac{\partial v}{\partial n}\right). \tag{4.39}$$

Comparing (4.39) with (3.31), it is seen that they can be reconciled because the restriction (3.32) implies that $\lambda = 1$.

Example 4.6. In connection with the biharmonic equation, one may consider the operator

$$\langle Pu, v\rangle = \int_\Omega v\Delta^2 u \, d\mathbf{x}. \tag{4.40}$$

Then

$$\langle (P-P^*)u, v\rangle = \int_{\partial\Omega} \left\{ v\frac{\partial \Delta u}{\partial n} - \Delta u \frac{\partial v}{\partial n} + \Delta v \frac{\partial u}{\partial n} - u\frac{\partial \Delta v}{\partial n}\right\} d\mathbf{x}. \tag{4.41}$$

In order to establish general canonical decompositions that are defined pointwise, one is led to consider the antisymmetric bilinear form $A : D \to D^*$, where $D = \mathfrak{R}^4$ and, for every $u = \{u_1, u_2, u_3, u_4\}$ and $v \in D$, one sets

$$\langle Au, v\rangle = v_1 u_3 + v_2 u_4 - v_3 u_1 - v_4 u_2. \tag{4.42}$$

Clearly the correspondence

$$\{u_1, u_2, u_3, u_4\} \to \{u, \Delta u, (\partial \Delta u/\partial n), (\partial u/\partial n)\}$$

transforms the expression (4.42) into that occurring in (4.41). Also

$$\{A\mathbf{u}, \mathbf{v}\rangle = \mathbf{v} \cdot \mathscr{A}\mathbf{u} \tag{4.43}$$

if the mapping $\mathscr{A} : D \to D$ is defined by

$$\mathscr{A}u = \{u_3, u_4, -u_1, -u_2\}. \tag{4.44}$$

Observe some properties that anticipate some general results that will be obtained later. These are:

(i) $\mathscr{A}^2 u = -u \quad \forall\, u \in D.$ (4.45)

(ii) If $I \subset D$ is regular, then

$$I \perp \mathscr{A}I. \tag{4.46}$$

(iii) $I \subset D$ is completely regular if and only if $\mathscr{A}I$ is the orthogonal complement of I.

Let four vectors $\mathbf{a}$, $\mathbf{b}$, $\mathbf{a}'$ and $\mathbf{b}'$, be linearly independent. Define

$$I_1=\{\mathbf{u}\in D \mid \mathbf{a}\cdot\mathbf{u}=\mathbf{b}\cdot\mathbf{u}=0\} \tag{4.47a}$$

and

$$I'=\{\mathbf{u}\in D \mid \mathbf{a}'\cdot\mathbf{u}=\mathbf{b}'\cdot\mathbf{u}=0\}. \tag{4.47b}$$

Condition (4.46) is satisfied by both subspaces I_1 and I_2, if and only if

$$\mathbf{a}\cdot\mathscr{A}\mathbf{b}=\mathbf{a}'\cdot\mathscr{A}\mathbf{b}'=0. \tag{4.48}$$

More explicitly, these conditions are

$$a_1b_3+a_2b_4-a_3b_1-a_4b_2=0 \tag{4.49a}$$

and

$$a_1'b_3'+a_2'b_4'+a_3'b_1'-a_4'b_2'=0. \tag{4.49b}$$

Arguments similar to those used in Example 4.5 allow the construction of general operators $B:D\to D^*$ that decompose $A:D\to D^*$ and general Green's formulas. A detailed discussion of this will be given in Chapter 6.

Example 4.7. All the examples of regular and completely regular subspaces that have been given thus far were defined pointwise. Let us consider now a subspace that will be defined by global conditions and which will be shown to be completely regular. Let $D=H^s(\Omega)$, $s\geqslant 2$, $N_P\subset D$ be the linear subspace of harmonic functions in Ω and define

$$I_P=\{u\in D \mid \exists\ w\in N_{P\ni}\ Au=Aw\}. \tag{4.50}$$

Here A is given by (3.16). The subspace $I_P\subset D$ is, therefore, the collection of functions which take the same relevant boundary values as harmonic functions in Ω. It will be assumed that the region Ω is two-dimensional, simply connected and with exterior E (Fig. 4.1). The fundamental solution of the Laplace equation is

$$G(\mathbf{x},\mathbf{y})=\frac{1}{2\pi}\ln|\mathbf{x}-\mathbf{y}|. \tag{4.51}$$

For every $\mathbf{y}\in E$, define

$$w_{\mathbf{y}}(\mathbf{x})=G(\mathbf{x},\mathbf{y}),\quad \mathbf{x}\in\Omega, \tag{4.52}$$

and let

$$\mathfrak{B}_E=\{w_{\mathbf{y}}(\mathbf{x}) \mid \mathbf{y}\in E\}. \tag{4.53}$$

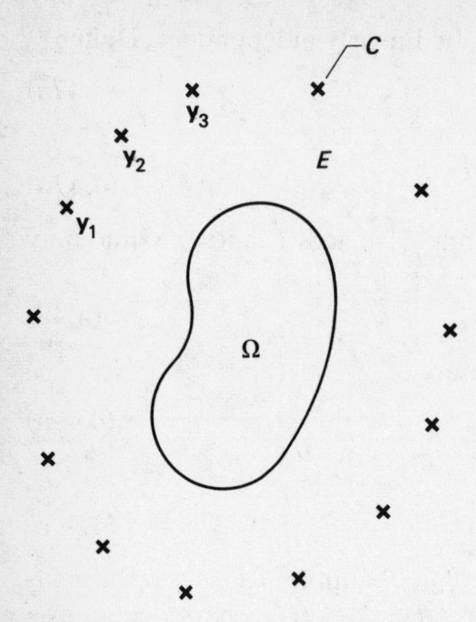

Fig. 4.1

Clearly $\mathfrak{B}_E \subset I_P$. If $\mathfrak{B}$ is replaced by $\mathfrak{B}_E$ and I by I_P, relation (4.10) becomes Green's third identity [94] and well-known results of potential theory [95] imply that $\mathfrak{B}_E \subset I_P$ is T-complete for I_P. In view of Lemma 4.3 this shows, by the way, that the subspace I_P is completely regular. This example anticipates general results to be shown later, according to which the linear subspaces of boundary values assumed by solution of homogeneous linear equations are, most frequently, completely regular.

Example 4.8. The system of functions $\mathfrak{B}_E$ given in Example 4.7 is non-denumerable. However, it is possible to construct a subset $\mathfrak{B} \subset \mathfrak{B}_E$ which is denumerable and more convenient for numerical applications. Indeed, let $\{\mathbf{y}_1, \mathbf{y}_2, \ldots\}$ be a dense subset of a curve C enclosing Ω (Fig. 4.1). Define

$$w_0 \equiv 1; \quad w_\alpha - w_{\mathbf{y}\alpha}, \quad \alpha = 1, 2, \ldots, \tag{4.54}$$

$$\mathfrak{B}_C = \{w_\mathbf{y} \mid \mathbf{y} \in C\} \tag{4.55}$$

and

$$\mathfrak{B} = \{w_0, w_1, w_2, \ldots\}. \tag{4.56}$$

Any harmonic function in Ω satisfies

$$\langle Au, w_0\rangle = \langle Au, 1\rangle = \int_{\partial\Omega} \frac{\partial u}{\partial n}\, d\mathbf{x} = 0. \tag{4.57}$$

The function

$$W(\mathbf{y}) = \langle Au, w_{\mathbf{y}}\rangle, \quad \mathbf{y} \in E \tag{4.58}$$

is harmonic on E. Christiansen [96] has shown that when (4.57) is satisfied the vanishing of $W(\mathbf{y})$ on C implies that this function vanishes identically on the exterior of C. Therefore, when this is the case, $W(\mathbf{y})$ vanishes identically on E, as can be seen by analytic continuation. This shows that $\{w_0\} \cup \mathfrak{B}_C$ is T-complete for I_P. Hence, $\mathfrak{B} \subset I_P$ is also T-complete, because

$$W(\mathbf{y}_\alpha) = \langle Au, w_\alpha\rangle = 0 \quad \forall\, \alpha = 1, 2, \ldots \tag{4.59}$$

implies that $W(\mathbf{y})$ vanishes identically on C, since $\{\mathbf{y}_1, \mathbf{y}_2, \ldots\} \subset C$ is dense in C.

Example 4.9. A similar argument can be used when Ω is the exterior of a simply connected and bounded region (Fig. 4.2). A convenient formulation of exterior boundary value problems for

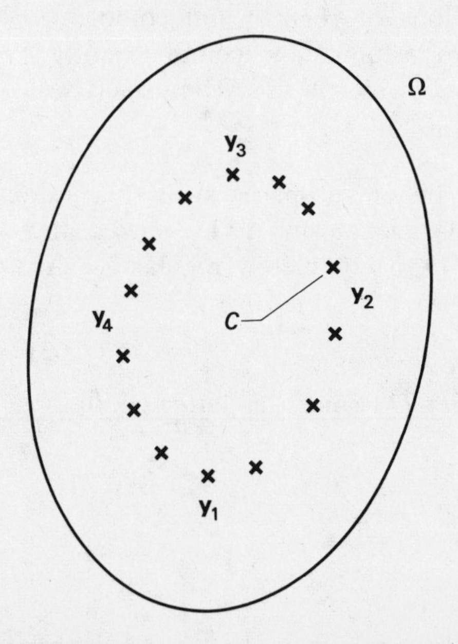

Fig. 4.2

the Laplace equation requires attention to be restricted to harmonic functions u such that $u - a_0 \log r$ go to zero when $r \to \infty$. Here [37]

$$a_0 = \frac{1}{2\pi} \int_{\partial\Omega} \frac{\partial u}{\partial n} \, d\mathbf{x}. \tag{4.60}$$

The curve C must be taken as the boundary of a simply connected region in the interior of $\partial\Omega$ and again $\{\mathbf{y}_1, \mathbf{y}_2, \ldots\} \subset C$ is a dense subset. In this case

$$\mathfrak{B} = \{w_1, w_2, \ldots\} \tag{4.61}$$

since, in general, (4.59) does not hold.

Remark 4.3. The structures discussed so far have two main ingredients: the linear space D and the antisymmetric bilinear functional $A: D \to D^*$. Assume that a second linear space D' and a bilinear functional $A': D' \to (D')^*$ are given. Let the mapping $\tau: D \to D'$ be onto D'; then τ is said to be a symplectic homomorphism between D, A and D', A', when

$$\langle Au, v\rangle = \langle A'\tau(u), \tau(v)\rangle. \tag{4.62}$$

It is a symplectic isomorphism if, in addition, τ is one-to-one.

It is easily seen that the notions of regular and completely regular subspaces, canonical decomposition and T-complete subsystem are invariant under symplectic isomorphisms. These facts will be used in later developments.

Remark 4.4. Let $D \supset D' \supset I$ be linear subspaces such that I is completely regular for $A: D \to D^*$. Denote by $A': D' \to (D')^*$ the restriction of A to D'. Then $I \subset D'$ is completely regular for A'. Indeed, for any $u \in D$ one has

$$\langle Au, v\rangle = 0 \quad \forall\, v \in I \Leftrightarrow u \in I. \tag{4.63}$$

Hence the same is true for any $u \in D'$, since this latter condition implies $u \in D$.

5 Green's formulas for operators defined in discontinuous fields

Problems formulated in discontinuous fields with prescribed jump conditions occur in many applications. In potential theory, for example, the jumps of the function and its normal derivative are usually prescribed, while in elasticity the data are the jumps of displacements and of tractions. Variational principles for this class of problems were developed by Prager [52]; Nemat-Nasser [53, 54] presents more recent surveys. The author has given an abstract formulation of such problems under the general heading of the problem of matching or connecting [22, 31, 35]. Here, general Green's formulas for such problems are presented which are applicable irrespective of the specific linear operators considered.

Consider two neighbouring regions R and E (Fig. 5.1) and let $\partial' R = \partial' E$ be the common boundary separating them; in addition, $\partial'' R$ and $\partial'' E$ will be the remaining parts of the boundaries of R and E, respectively. Let D_R and D_E be two linear spaces; in the applications to be made their elements will be functions defined on R and on E, respectively. Consider the product space $\hat{D} = D_R \oplus D_E$; elements $\hat{u} \in \hat{D}$ are pairs $\hat{u} = \{u_R, u_E\}$ where $u_R \in D_R$ while $u_E \in D_E$. Given operators $P_R : D_R \to D_R^*$ and $P_E : D_E \to D_E^*$, define $\hat{P} : \hat{D} \to \hat{D}^*$ by

$$\langle \hat{P}\hat{u}, \hat{v}\rangle = \langle P_R u_R, v_R\rangle + \langle P_E u_E, v_E\rangle. \tag{5.1}$$

Example 5.1. Let Ω be a bounded and simply connected region of $\mathfrak{R}^n$ (Fig. 5.1). Assume, further, that the regions $R \subset \Omega$ and $E \subset \Omega$ constitute a partition of Ω; i.e.,

$$\bar{R} \cup \bar{E} = \bar{\Omega} \quad \text{while} \quad R \cap E = \phi. \tag{5.2}$$

Here, the bar refers to the closure of the corresponding region. Let $s \geqslant 2$ be fixed. Define D_R as the linear space whose elements are the restrictions to R, of functions belonging to $H^s(R)$. The linear space D_E is defined replacing R by E. Define $P_R : D_R \to D_R^*$

$$\langle P_R u_R, v_R\rangle = \int_R v \Delta u \, d\mathbf{x} + \int_{\partial'' R} u \frac{\partial v}{\partial n} d\mathbf{x} \tag{5.3}$$

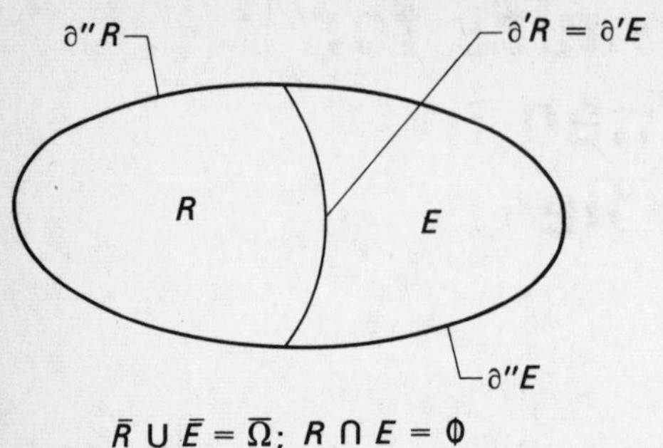

Fig. 5.1

and $P_E : D_E \to D_E^*$, replacing R by E above. Then

$$\langle \hat{P}\hat{u}, \hat{v}\rangle = \int_{R\cup E} v\Delta u \, d\mathbf{x} + \int_{\partial''(R\cup E)} u \frac{\partial v}{\partial n} \, d\mathbf{x}. \tag{5.4}$$

Here, the subindexes R and E have been deleted from the functions u and v, because they are specified by the index under the integral sign. Also, the notation

$$\partial''(R\cup E) = \partial''R \cup \partial''E \tag{5.5}$$

has been introduced and the relation

$$\int_{R\cup E} v\Delta u \, d\mathbf{x} = \int_R v\Delta u \, d\mathbf{x} + \int_E v\Delta u \, d\mathbf{x} \tag{5.6}$$

is used.

Let $\hat{A} = \hat{P} - \hat{P}^*$, $A_R = P_R - P_R^*$ and $A_E = P_E - P_E^*$. From (5.1) it follows that

$$\langle \hat{A}\hat{u}, \hat{v}\rangle = \langle A_R u_R, v_R\rangle + \langle A_E u_E, v_E\rangle. \tag{5.7}$$

In order to simplify the notation a little, the symbols $\hat{N}_A$, N_{AR} and N_{AE} will be used to denote the null subspaces of $\hat{A}$, A_R and A_E, respectively. The relation

$$\hat{N}_A = N_{AR} \oplus N_{AE} \tag{5.8}$$

will be used later; it is equivalent to

$$\hat{u} = \{u_R, u_E\} \in \hat{N}_A \Leftrightarrow u_R \in N_{AR} \quad \text{and} \quad u_E \in N_{AE}. \tag{5.9}$$

This latter relation follows from (5.7).

Example 5.2. In the previous example, $A_R : D_R \to D_R^*$ is given by

$$\langle A_R u_R, v_R \rangle = \int_{\partial' R} \left\{ v \frac{\partial u}{\partial n} - u \frac{\partial v}{\partial n} \right\} d\mathbf{x}, \tag{5.10}$$

where the unit normal vector $\mathbf{n}$ points outwards from R. In this connection a note of warning must be given to the reader. Two different unit normal vectors will be used there, one pointing outwards from R and another one pointing outwards from E. These will be denoted by $\mathbf{n}_R$ and $\mathbf{n}_E$, respectively. When writing integrals on $\partial' R = \partial' E$ the sense of the normal vector is given by the index under the integral sign, unless explicitly specified otherwise. Notice that the symbols u and v in equation (5.10) may be ambiguous, because, given $\hat{u} = \{u_R, u_E\}$, one may consider on $\partial' R = \partial' E$ either the continuous extension of u_R or that of u_E. As already mentioned, the function to be used is specified by the index under the integral sign unless something else is explicitly specified. Thus, for example, (5.10) in a more explicit notation is

$$\langle A_R u_R, v_R \rangle = \int_{\partial' R} \left\{ v_R \frac{\partial u_R}{\partial n_R} - u_R \frac{\partial v_R}{\partial n_R} \right\} d\mathbf{x}. \tag{5.11}$$

As a further illustration, observe that when these conventions are used, the following relation holds:

$$\int_{\partial' R} \left\{ v \frac{\partial u}{\partial n} - u \frac{\partial v}{\partial n} \right\} d\mathbf{x} = - \int_{\partial' E} \left\{ v_R \frac{\partial u_R}{\partial n} - u_R \frac{\partial v_R}{\partial n} \right\} d\mathbf{x}, \tag{5.12}$$

since the normal vectors used in these integrals have opposite senses.

Replacing R by E, keeping in mind the conventions just mentioned, one obtains $A_E : D_E \to D_E^*$. The operator $\hat{A} : \hat{D} \to \hat{D}^*$ is given by

$$\langle \hat{A}\hat{u}, \hat{v} \rangle = \int_{\partial' R} \left\{ v \frac{\partial u}{\partial n} - u \frac{\partial v}{\partial n} \right) d\mathbf{x} + \int_{\partial' E} \left\{ v \frac{\partial u}{\partial n} - u \frac{\partial v}{\partial n} \right\} d\mathbf{x}. \tag{5.13}$$

A less symmetric but useful expression is

$$\langle \hat{A}\hat{u}, \hat{v} \rangle = \int_{\partial' R} \left\{ \left(v \frac{\partial u}{\partial n} - v_E \frac{\partial u_E}{\partial n} \right) - \left(u \frac{\partial v}{\partial n} - u_E \frac{\partial v_E}{\partial n} \right) \right\} d\mathbf{x}. \tag{5.14}$$

Using equation (5.10) it is easy to see that

$$N_{AR} = \left\{ u_R \in D_R \mid u_R = \frac{\partial u_R}{\partial n} = 0, \quad \text{on} \quad \partial' R \right\}. \tag{5.15}$$

Hence a corresponding relation holds for N_{AE}. Application of (5.8) yields

$$\hat{N}_A = \left\{ \hat{u} \in \hat{D} \mid u_R = u_E = \frac{\partial u_R}{\partial n} = \frac{\partial u_E}{\partial n} = 0, \quad \text{on} \quad \partial' R \right\}. \tag{5.16}$$

In the general development of the theory a linear subspace $\hat{S} \subset \hat{D}$ will be considered. Elements $\hat{u} = \{u_R, u_E\} \in \hat{S}$ will be said to be smooth. When $\hat{u} = \{u_R, u_E\}$ is smooth, $u_R \in D_R$ and $u_E \in D_E$ will be said to be smooth extensions of each other.

Definition 5.1. *Let $\hat{S} \subset \hat{D} = D_R \oplus D_E$ be a linear subspace. Then $\hat{S}$ is said to be a smoothness relation if every $u_R \in D_R$ possesses at least one smooth extension $u_E \in D_E$ and conversely.*

A smoothness relation $\hat{S}$ is said to be regular or completely regular for $\hat{P}$, when as a subspace it is regular or completely regular for $\hat{P}$, respectively. Therefore, a smoothness relation $\hat{S}$ is regular when

$$\text{(a)} \quad \hat{S} \supset \hat{N}_A \tag{5.17a}$$

and

$$\text{(b)} \quad \langle \hat{A}\hat{u}, \hat{v} \rangle = 0 \quad \forall\, \hat{u} \in \hat{S} \quad \& \quad \hat{v} \in \hat{S}. \tag{5.17b}$$

Similarly, it is completely regular when

$$\langle \hat{A}\hat{u}, \hat{v} \rangle = \langle A_R u_R, v_R \rangle + \langle A_E u_E, v_E \rangle = 0 \quad \forall\, \hat{v} \in \hat{S} \Leftrightarrow \hat{u} \in \hat{S}. \tag{5.18}$$

Example 5.3. This is a continuation of Examples 5.1 and 5.2. The linear subspace of smooth functions $\hat{S} \subset \hat{D} = D_R \oplus D_E$ is defined by

$$\hat{S} = \left\{ \hat{u} \in \hat{D} \mid u_R = u_E \quad \text{and} \quad \frac{\partial u_R}{\partial n_R} = \frac{\partial u_E}{\partial n_R}, \quad \text{on} \quad \partial' R = \partial' E \right\}. \tag{5.19}$$

Clearly, $\hat{S}$ is a smoothness relation since it satisfies the conditions of Definition 5.1. Moreover, $\hat{S}$ is regular. Indeed, (5.17a) is satisfied by virtue of (5.16) and (5.19). Also, when $\hat{u} \in \hat{S}$ and $\hat{v} \in \hat{S}$, equation (5.14) reduces to

$$\langle A\hat{u}, \hat{v} \rangle = 0 \tag{5.20}$$

by virtue of (5.19). Thus, $\hat{S} \subset \hat{D}$ is a regular smoothness relation.

Actually, it can be shown that $\hat{S}$ is completely regular. However, this will follow from Corollary 5.1.

The mapping $\tau:\hat{D}\to\hat{D}$, defined for every $\hat{u}=\{u_R, u_E\}\in\hat{D}$ by

$$\tau\hat{u}=\{u_R, -u_E\}, \tag{5.21}$$

will be used in the following discussion. Observe that τ is self-inverse; i.e.,

$$\tau^2=1. \tag{5.22}$$

Note also that

$$\tau\hat{N}_A=\hat{N}_A \tag{5.23}$$

and

$$\langle\hat{A}\hat{u}, \hat{v}\rangle=\langle\hat{A}\tau\hat{u}, \tau\hat{v}\rangle \tag{5.24}$$

by virtue of (5.8) and (5.7), respectively. Thus, τ is a symplectic isomorphism.

When a smoothness relation $\hat{S}\subset\hat{D}$ is given, one can always associate with it another linear subspace which, together with $\hat{S}$, spans the space $\hat{D}$.

Definition 5.2. *Given a smoothness relation $\hat{S}\subset\hat{D}$, the linear subspace*

$$\hat{M}=\tau\hat{S}=\{\hat{u}\in\hat{D}\mid\tau\hat{u}\in\hat{S}\} \tag{5.25}$$

will be associated to it. Elements $\hat{u}=\{u_R, u_E\}$ belonging to $\hat{M}$ are said to have zero mean.

Observe that the relation $\tau\hat{S}=\tau^{-1}\hat{S}$ was used when writing (5.25). Also, as already mentioned, the space $\hat{M}$ has the property that

$$\hat{D}=\hat{S}+\hat{M} \tag{5.26}$$

because any element $\hat{u}=\{u_R, u_E\}\in\hat{D}$ can be written as

$$\hat{u}=\dot{u}-\tfrac{1}{2}[\hat{u}] \tag{5.27}$$

where $\dot{u}\in\hat{S}$ and $[\hat{u}]\in\hat{M}$ are given by

$$\dot{u}=\tfrac{1}{2}\{u'_R+u_R, u'_E+u_E\} \tag{5.28a}$$

and

$$[\hat{u}]=\{u'_R-u_R, u'_E-u_E\}. \tag{5.28b}$$

Here $u'_R \in D_R$ is a smooth extension of $u_E \in D_E$. Conversely, $u'_E \in D_E$ is a smooth extension of $u_R \in D_R$.

The previous construction leads in a straightforward manner to a canonical decomposition of $\hat{D}$ with respect to $\hat{P}: \hat{D} \to \hat{D}^*$.

Theorem 5.1. *When the smoothness relation $\hat{S}$ is regular, the pair $\{\hat{S}, \hat{M}\}$ is a canonical decomposition of $\hat{D}$, with respect to $\hat{P}: \hat{D} \to \hat{D}^*$.*

Proof. In view of Definition 4.5 and (5.26), it remains to prove that the regularity of the subspace $\hat{S}$ implies the regularity of $\hat{M}$; i.e., that relations (5.17) hold for $\hat{M}$ whenever they hold for $\hat{S}$. Clearly

$$\hat{S} \supset \hat{N}_A \Rightarrow \hat{M} = \tau\hat{S} \supset \tau\hat{N}_A = \hat{N}_A \tag{5.29a}$$

by virtue of (5.23). Also,

$$\langle \hat{A}\hat{u}, \hat{v} \rangle = 0, \quad \hat{u} \in \hat{S} \quad \& \quad \hat{v} \in \hat{S} \Rightarrow \langle A\hat{u}, \hat{v} \rangle = 0 \qquad \forall\, \hat{u} \in \hat{M} \quad \& \quad \hat{v} \in \hat{M}, \tag{5.29b}$$

in view of (5.24) and (5.25).

Corollary 5.1. *Every regular smoothness relation $\hat{S}$ is completely regular.*

Proof. By Theorem 4.1, since the construction of $\hat{M}$ is always possible.

Corollary 5.2. *When a smoothness relation $\hat{S}$ is regular the continuous extension $u'_E \in D_E$ of any $u_R \in D_R$ is unique except for elements of $N_{AE} \subset D_E$. The dual result is also true.*

Proof. Let $\{u_R, u'_E\} \in \hat{S}$ and $\{u_R, u''_E\} \in \hat{S}$. Then $\{0, u'_E - u''_E\} \in \hat{S}$ and also $\{0, -(u'_E - u''_E)\} \in \hat{S}$. Thus $\{0, u'_E - u''_E\} \in \hat{S} \cap \hat{M} = \hat{N}_A$. Hence $u'_E - u''_E \in N_{AE}$ by virtue of (5.8).

Before we proceed further, it is convenient to make the observation that, according to Theorem 5.1, the mapping $\tau: \hat{D} \to \hat{D}^*$ defined by (5.21) has permitted us to construct the set $\hat{M} = \tau\hat{S}$, which, together with $\hat{S}$, constitutes a canonical decomposition $\hat{D}$. However, the set $\hat{M}$ is only one of the many possible choices one can make.

Example 5.4. Applying Definition (5.25) when $\hat{S}$ is given by (5.19), one obtains

$$\hat{M} = \left\{ \hat{u} \in \hat{D} \mid u_R + u_E = 0 \text{ and } \frac{\partial u_R}{\partial n_R} + \frac{\partial u_E}{\partial n_R} = 0, \quad \text{on} \quad \partial' R = \partial' E \right\}, \tag{5.30}$$

i.e., the averages of the function and of its normal derivative vanish across the surface of discontinuity. According to the nomenclature already introduced, elements having this property are said to have zero mean.

The following observation will permit simplification of computations in specific applications.

Corollary 5.3. *When the smoothness relation* $\hat{S}$ *is regular, one has*

$$\langle \hat{A}\hat{u}, \hat{v} \rangle = 2\langle A_R u_R, v_R \} = 2\langle A_E u_E, v_E \rangle, \quad \forall\, \hat{u} \in \hat{M} \,\&\, \hat{v} \in \hat{S}. \tag{5.31}$$

Here $\hat{u} = \{u_R, u_E\}$ *and* $\hat{v} = \{v_R, v_E\}$.

Proof. By Definition (5.25), one has $\{u_R, -u_E\} \in \hat{S}$. Hence

$$\langle A_R u_R, v_R \rangle = \langle A_E u_E, v_E \rangle \tag{5.32}$$

by virtue of (5.18). Substitution of equation (5.32) into (5.7) yields the first of equations (5.31). The second of these equations can be obtained by duality.

In general, when a canonical decomposition is available, it is possible to construct Green's formulas by applying equation (4.23). Indeed, any element $\hat{u} \in \hat{D}$ can be written as

$$\hat{u} = \hat{u}_1 + \hat{u}_2, \quad \hat{u}_1 \in \hat{S} \,\&\, \hat{u}_2 \in \hat{M}. \tag{5.33}$$

According to (5.27),

$$\hat{u}_1 = \dot{u}; \qquad \hat{u}_2 = -\tfrac{1}{2}[\hat{u}]. \tag{5.34}$$

However, the elements $\dot{u} \in \hat{S}$ and $[\hat{u}] \in \hat{M}$, are not uniquely defined. They are defined up to elements of $\hat{N}_A$, i.e., they belong to unique cosets of the space $\hat{D}/\hat{N}_A$. In what follows, elements $\dot{u}$ and $[\hat{u}]$ satisfying equations (5.28) will be called the mean and the jump of $\hat{u}$, respectively.

By means of Theorem 4.2 it is now possible to define an operator that decomposes $\hat{A} : \hat{D} \to \hat{D}^*$; this will be denoted by $\hat{J} : \hat{D} \to \hat{D}^*$. According to (4.23) and (5.34), it satisfies

$$2\langle \hat{J}\hat{u}, \hat{v} \rangle = 2\langle \hat{A}\hat{u}_2, \hat{v}_1 \rangle = -\langle \hat{A}[\hat{u}], \dot{v} \rangle. \tag{5.35}$$

This operator will be called the jump operator; this nomenclature is motivated by the fact that

$$\hat{J}\hat{u}=0 \Leftrightarrow [\hat{u}] \in \hat{N}_A \Leftrightarrow \hat{u} \in \hat{S}, \tag{5.36}$$

as can be easily verified. Application of Corollary 5.3 to equation (5.35) yields

$$\langle \hat{J}\hat{u}, \hat{v}\rangle = -\langle A_R[\hat{u}]_R, (\dot{v})_R\rangle, \tag{5.37a}$$

where

$$[\hat{u}]_R = u'_R - u_R, \qquad (\dot{v})_R = \tfrac{1}{2}(v'_R + v_R) \tag{5.37b}$$

by virtue of equations (5.28). Recall that u'_R and v'_R are continuous extensions of u_E and v_E, respectively. The dual formula is

$$\langle \hat{J}\hat{u}, \hat{v}\rangle = -\langle A_E[\hat{u}]_E, (\dot{v})_E\rangle \tag{5.38a}$$

with

$$[\hat{u}]_E = u'_E - u_E, \qquad (\dot{v})_E = \tfrac{1}{2}(v'_E + v_E), \tag{5.38b}$$

where u'_E and v'_E are continuous extensions of u_R and v_R, respectively.

In view of Theorem 4.2, our results can be summarized by means of the theorem that follows.

Theorem 5.2. *Assume that* $\hat{S} \subset \hat{D}$ *is a regular smoothness relation for* $\hat{P}: \hat{D} \to \hat{D}^*$. *Define* $\hat{J}: \hat{D} \to \hat{D}^*$ *by means of* (5.37) *or, equivalently, by* (5.38). *Then*

$$\hat{A} = \hat{P} - \hat{P}^* = \hat{J} - \hat{J}^* \tag{5.39}$$

is a Green's formula.

Proof. The previous discussion.

Example 5.5. Let us obtain a Green's formula for Laplace's operator in discontinuous fields. A specific application of such formulas is the derivation of variational principles such as those obtained by Prager [52]. To this end, apply equations (5.38) using (5.11). Then

$$\langle \hat{J}\hat{u}, \hat{v}\rangle = -\int_{\partial R} \left\{ \dot{v}\frac{\partial[\hat{u}]}{\partial n} - [\hat{u}]\frac{\partial \dot{v}}{\partial n} \right\} d\mathbf{x}, \tag{5.40}$$

where $(\dot{v})_R$ and $[\hat{u}]_R$ are given by (5.37b). Taking into account the smoothness condition (5.19), equation (5.40) can be applied with

$$[\hat{u}]_R = u_E - u_R; \quad \frac{\partial[\hat{u}]_R}{\partial n} = \frac{\partial u_E}{\partial n} - \frac{\partial u_R}{\partial n}, \quad \text{on} \quad \partial' R \tag{5.41a}$$

and

$$(\dot{v})_R = \tfrac{1}{2}(v_E + v_R); \quad \frac{\partial(\dot{v})_R}{\partial n} = \frac{1}{2}\left(\frac{\partial v_E}{\partial n} + \frac{\partial v_R}{\partial n}\right), \quad \text{on} \quad \partial' R. \tag{5.41b}$$

Observe that $[\hat{u}]_R$ and $\partial[u]_R/\partial n$ give the jumps of the function and of its normal derivative when going from R into E across the surface of discontinuity. Also, $(\dot{v})_R$ and $\partial(\dot{v})_R/\partial n$ are the average values of the function and of its normal derivatives across the discontinuity. The corresponding Green's formula is

$$\int_{R\cup E} \{v\Delta u - u\Delta v\}\, d\mathbf{x} + \int_{\partial''(R\cup E)} \left\{u\frac{\partial v}{\partial n} - v\frac{\partial u}{\partial n}\right\} d\mathbf{x}$$
$$= \int_{\partial' R} \left\{[\hat{u}]\frac{\partial \dot{v}}{\partial n} - \dot{v}\frac{\partial[\hat{u}]}{\partial n}\right\} d\mathbf{x} - \int_{\partial' R} \left\{[\hat{v}]\frac{\partial \dot{u}}{\partial n} - \dot{u}\frac{\partial[\hat{v}]}{\partial n}\right\} d\mathbf{x}. \tag{5.42}$$

Of course this relation still holds if R and E are interchanged, recalling that when so doing the normal vector on $\partial' R = \partial' E$ changes its sense.

Example 5.6. Let k_R and k_E be two real numbers. Then the previous discussion can be modified as follows. Take D_R and D_E as in Example 5.3. Define $P_R : D_R \to D_R^*$ by

$$\langle P_R u_R, v_R\rangle = k_R\left\{\int_R v\Delta u\, d\mathbf{x} + \int_{\partial'' R} u\frac{\partial v}{\partial n}\right\} d\mathbf{x} \tag{5.43}$$

and $P_E : D_E \to D_E^*$, replacing R by E. Then

$$\langle A_R u_R, v_R\rangle = k_R\int_{\partial' R} \left\{v\frac{\partial u}{\partial n} - u\frac{\partial v}{\partial n}\right\} d\mathbf{x}. \tag{5.44}$$

Equation (5.13) becomes

$$\langle \hat{A}\hat{u}, \hat{v}\rangle = k_R\int_{\partial' R} \left\{v\frac{\partial u}{\partial n} - u\frac{\partial v}{\partial n}\right\} d\mathbf{x} + k_E\int_{\partial' E} \left\{v\frac{\partial u}{\partial n} - u\frac{\partial v}{\partial n}\right\} d\mathbf{x}. \tag{5.45}$$

Corresponding to (5.14), we have

$$\langle \hat{A}\hat{u}, \hat{v}\rangle = \int_{\partial' R} \left\{\left(v_R k_R\frac{\partial u_R}{\partial n} - v_E k_E\frac{\partial u_E}{\partial n}\right) - \left(u_R k_R\frac{\partial v_R}{\partial n} - u_E k_E\frac{\partial v_E}{\partial n}\right)\right\} d\mathbf{x}. \tag{5.46}$$

Equation (5.16) remains valid.

Let us define the smoothness relation $\hat{S} \subset \hat{D} = D_R \oplus D_E$, by

$$\hat{S} = \left\{ \hat{u} \in \hat{D} \mid u_R = u_E \quad \text{and} \quad k_R \frac{\partial u_R}{\partial n_R} = k_E \frac{\partial u_E}{\partial n_R}, \quad \text{on} \quad \partial' R = \partial' E \right\}. \tag{5.47}$$

In applications to flow through porous media, $\hat{S}$ is the space of functions satisfying the usual conditions of continuity of the pressure (or piezometric head) and the flux. It can be seen that $\hat{S}$ is a smoothness relation because the conditions of Definition 5.1 are satisfied. Moreover, $\hat{S}$ is regular, because firstly $\hat{S} \supset \hat{N}_A$, as can be verified by comparing (5.16) and (5.47). Secondly, when $\hat{u} \in \hat{S}$ and $\hat{v} \in \hat{S}$, equation (5.46) reduces to $\langle \hat{A}\hat{u}, \hat{v} \rangle = 0$, by virtue of (5.47). Thus, equations (5.38) can be applied to obtain a Green's formula. Using (5.44), this yields

$$\langle \hat{J}\hat{u}, \hat{v} \rangle = -k_R \int_{\partial' R} \left\{ \dot{v} \frac{\partial [u]}{\partial n} - [\hat{u}] \frac{\partial \dot{v}}{\partial n} \right\} \mathrm{d}\mathbf{x} \tag{5.48}$$

with

$$[u]_R = u'_R - u_R; \quad \frac{\partial [u]}{\partial n} = \frac{\partial u'_R}{\partial n} - \frac{\partial u_R}{\partial n}, \quad \text{on} \quad \partial' R \tag{5.49a}$$

$$(\dot{v})_R = \tfrac{1}{2}(u'_R + u_R); \quad \frac{\partial (\dot{v})_R}{\partial n} = \frac{1}{2}\left(\frac{\partial u'_R}{\partial n} + \frac{\partial u_R}{\partial n} \right), \quad \text{on} \quad \partial' R. \tag{5.49b}$$

Now the continuity conditions (5.47) imply that

$$u'_R = u_E \quad \text{and} \quad \frac{\partial u'_R}{\partial n} = \frac{k_E}{k_R} \frac{\partial u_E}{\partial n}, \quad \text{on} \quad \partial' R \tag{5.50}$$

and similarly for v'_R. Substitution of (5.49) and (5.50) into (5.48), yields

$$\langle \hat{J}\hat{u}, \hat{v} \rangle = \int_{\partial' R} \left\{ [u] \left(k \frac{\partial v}{\partial n} \right)_a - \dot{v} \left[k \frac{\partial u}{\partial n} \right] \right\} \mathrm{d}\mathbf{x}, \tag{5.51}$$

where we have written

$$\begin{aligned} \left[k \frac{\partial u}{\partial n} \right]_R &= k_E \frac{\partial u_E}{\partial n} - k_R \frac{\partial u_R}{\partial n}; \\ \left(k \frac{\partial \dot{v}}{\partial \dot{n}} \right)_a &= \frac{1}{2} \left(k_E \frac{\partial v_E}{\partial n} + k_R \frac{\partial v_R}{\partial n} \right) \quad \text{on} \quad \partial' R. \end{aligned} \tag{5.52}$$

The corresponding Green's formula is given by (5.39) and we dispense with writing it in full since it is straightforward.

6 Illustrations of Green's formulas

In this chapter general examples of Green's formulas are presented. Many of the operators listed are formally symmetric in the classical sense; others can be included due to the extension of this concept implied by Definitions 3.1 to 3.3, which yield the criterion contained in Theorem 3.1.

6.1 Elliptic equations

The most general elliptic operator of order $2m$ that is self-adjoint is of the form given by equations (3.18) and (3.19). Let C_j, $0 \leq j \leq \nu - 1$, be ν differential operators defined on the boundary $\partial\Omega$ by

$$C_j u = \sum_{|h| \leq m_j} c_{jh}(x) D^h u, \tag{6.1}$$

where c_{jh} are infinitely differentiable in $\partial\Omega$ and m_j is the order of C_j.

The system of operators $\{C_j\}_{j=0}^{\nu-1}$ is a normal system on $\partial\Omega$ if [82]:

(a) $\sum_{|h|=m_j} C_{jh}(\mathbf{x})\boldsymbol{\xi}^h \neq 0 \quad \forall \mathbf{x} \in \partial\Omega$, and

$$\forall \boldsymbol{\xi} \neq 0 \text{ which is normal to } \partial\Omega, \tag{6.2a}$$

(b) $m_j \neq m_i$, for $j \neq i$. (6.2b)

In addition, let $\{F_i\}_{i=0}^{\nu-1}$ be a system of ν differential operators defined on $\partial\Omega$ by

$$F_i u = \sum_{|h| \leq m_i} f_{ih}(x) D^h_u. \tag{6.3}$$

The system $\{F_i\}_{i=0}^{\nu-1}$ is a Dirichlet system of order ν on $\partial\Omega$, if it is normal on $\partial\Omega$ and if the orders m_i run through exactly $0, 1, \ldots, \nu-1$ when i goes from 0 to $\nu-1$.

A theorem on Green's formulas for the case when the differential operator $\mathscr{L}$ of order $2m$, defined by equation (3.18), is formally symmetric and elliptic is the following. Let $\{C_j\}_{j=0}^{m-1}$ be a normal system on $\partial\Omega$, given by (6.1), with $m_j \leq 2m-1$. It is always possible to choose, non-uniquely, another system of boundary operators $\{S_j\}_{j=0}^{m-1}$ normal on $\partial\Omega$ and of orders $\mu_j \leq 2m-1$, such that the system $\{C_0, \ldots, C_{m-1}, S_0, \ldots, S_{m-1}\}$ is a Dirichlet system of order $2m$ on $\partial\Omega$ and the following Green's formula holds [82]:

$$\int_\Omega v\mathcal{L}u \, d\mathbf{x} - \int_\Omega u\mathcal{L}v \, d\mathbf{x} = \sum_{j=0}^{m-1} \int_{\partial\Omega} S_j u C_j v \, d\mathbf{x} - \sum_{j=0}^{m-1} \int_{\partial\Omega} S_j v C_j u \, d\mathbf{x} \tag{6.4}$$

for every $u \in D$ and $v \in D$, where D is taken as in Example 3.4.

Let $P: D \to D^*$ be defined by (3.19) and $A = P - P^*$. Set $D = H^s(\Omega)$; $s \geq 2m$. Define $B: S \to D^*$ by

$$\langle Bu, v\rangle = \sum_{j=0}^{m-1} \int_{\partial\Omega} S_j u C_j v \, d\mathbf{x}, \tag{6.5}$$

where C_j is of order $m_j \leq 2m-1$, while S_j is of order $\mu_j \leq 2m-1$. Assume $\{C_0, \ldots, C_{m-1}, S_0, \ldots, S_{m-1}\}$ is a Dirichlet system of order $2m$ on $\partial\Omega$ and Green's formula (6.4) holds. Clearly, $A = B - B^*$ and it can be shown that B and B^* can be varied independently; therefore, B decomposes A and the relation $A = B - B^*$ is a Green's formula in the sense of Definition 3.4. This is easy to see because conditions (6.2) imply that, given $U \in D$ and $V \in D$, one can find $u \in D$ such that

$$C_0 U = C_0 u, \ldots, C_{m-1} U = C_{m-1} u \quad \text{while} \quad S_0 V = S_0 u, \ldots, S_{m-1} V = S_{m-1} u. \tag{6.6}$$

Indeed, if equations (6.6) are reordered, listing them in an increasing order for the operators involved, then $u \in D$ can be constructed so that their normal derivatives on $\partial\Omega$ satisfy

$$a_j \frac{\partial^j u}{\partial n^j} = L_j\left(\frac{\partial^j U}{\partial n}, \frac{\partial^{j-1}(U-u)}{\partial n^{j-1}}, \ldots, V-u\right), \quad j = 1, \ldots, 2m-1, \tag{6.7a}$$

and simultaneously

$$a_0 u = L_0(U). \tag{6.7b}$$

Here L_j $(j = 0, 1, \ldots, 2m-1)$ are linear functionals of the specified arguments, while

$$a_j = \sum_{|\mathbf{h}|=j} f_{j\mathbf{h}}(\mathbf{x})\mathbf{n}^{\mathbf{h}} \neq 0. \tag{6.8}$$

Here, **n** is the unit normal vector to $\partial\Omega$. Clearly, the system (6.7) can be solved recursively.

The converse, however, is not true. As an illustration, in Example 3.8,

$$Cu = a_1 u + a_2 \frac{\partial u}{\partial n}; \qquad Su = b_1 u + b_2 \frac{\partial u}{\partial n}, \tag{6.9}$$

where

$$a_1 b_2 - a_2 b_1 = 1. \tag{6.10}$$

The system $\{C, S\}$ is not a Dirichlet system of order 2, whenever $a_2 \neq 0$ and simultaneously $b_2 \neq 0$.

Thus, the concept of Green's formula of Definition 3.4 is wider than the one usually considered in the theory of partial differential equations [82]. It makes it possible, for example, to carry out a systematic search for positive definite and symmetric operators associated with boundary value problems—a question which is relevant for the formulation of extremal variational principles. Consider the equation

$$Pu - Bu = PU - BV, \tag{6.11}$$

with $P: D \to D^*$ given by (2.8) and $B: D \to D^*$ by (3.30), subjected to (3.31). Now $P - B$ is a symmetric operator and

$$\langle Pu, u\rangle = \int_\Omega \nabla u \cdot \nabla u \, d\mathbf{x} - \int_{\partial\Omega} \left\{ a_1 b_1 u^2 + a_2 b_2 \left(\frac{\partial u}{\partial n}\right)^2 + (a_1 b_2 + a_2 b_1 + 1) u \frac{\partial u}{\partial n} \right\} d\mathbf{x}. \tag{6.12}$$

Taking into account restriction (3.31), this is non-negative if and only if $a_1 b_1 \leqslant 0$, $a_2 b_2 \leqslant 0$ and simultaneously

$$a_1^2 b_2^2 \leqslant a_1 b_2 a_2 b_1 = a_1 b_2 (a_1 b_2 - 1), \tag{6.13}$$

i.e.

$$a_1 b_2 \leqslant 0. \tag{6.14}$$

This implies that $a_2 \neq 0$, because otherwise (6.14) is incompatible with (3.31). When $a_2 \neq 0$,

$$b_1 = \frac{a_1 b_2 - 1}{a_2}. \tag{6.15}$$

Hence

$$\frac{a_1}{a_2} \geqslant 0; \qquad a_2 b_2 \leqslant 0. \tag{6.16}$$

Notice that these two inequalities imply (6.14).

A more complicated situation occurs when considering the biharmonic equation. In the discussion of Green's formulas for this case, which are defined pointwise, one is led to consider (Example 4.6) a bilinear functional $A: \hat{\mathfrak{H}} \to \hat{\mathfrak{H}}^*$, where $\hat{\mathfrak{H}} = \mathfrak{H} \oplus \mathfrak{H}$ and $\mathfrak{H} = \mathfrak{R}^2$. This is a Hilbert space when it is provided with the l_2, structure. Recall equations (4.47); they are

$$I_1 = \{\mathbf{u} \in D \mid \mathbf{a} \cdot \mathbf{u} = \mathbf{b} \cdot \mathbf{u} = 0\} \tag{6.17a}$$

and

$$I' = \{\mathbf{u} \in D \mid \mathbf{a}' \cdot \mathbf{u} = \mathbf{b}' \cdot \mathbf{u} = 0\}. \tag{6.17b}$$

Without lack of generality it can be assumed that

$$\mathbf{a} \perp \mathbf{b}, \mathbf{a}' \perp \mathbf{b}', \|\mathbf{a}\| = \|\mathbf{b}\| = \|\mathbf{a}'\| = 1, \tag{6.18}$$

together with $(\mathbf{a}, \mathscr{A}\mathbf{b}) = (\mathbf{a}', \mathscr{A}\mathbf{b}') = 0$ as required by (4.48). Here, $\mathscr{A}: \hat{\mathfrak{H}} \to \hat{\mathfrak{H}}$ is given by (4.44). With this definition

$$\langle A\mathbf{u}, \mathbf{v}\rangle = (\mathscr{A}\mathbf{u}, \mathbf{v}). \tag{6.19}$$

Clearly, the representation

$$\mathbf{u} = \mathbf{u}_1 + \mathbf{u}'; \qquad \mathbf{u}_1 \in I_1, \mathbf{u}' \in I' \tag{6.20}$$

applies. In view of (6.20) and (4.23),

$$\langle B\mathbf{u}, \mathbf{v}\rangle = (\mathbf{v}_1, \mathbf{u}_2) \tag{6.21}$$

decomposes A. Here

$$\mathbf{u}_2 = \operatorname{proj} \mathscr{A}\mathbf{u}' \tag{6.22}$$

and the projection is taken on I_1. In addition

$$\mathbf{v}_1 = \alpha \mathscr{A}\mathbf{a} + \beta \mathscr{A}\mathbf{b}, \tag{6.23}$$

where the real numbers α and β satisfy

$$\alpha(\mathbf{a}', \mathscr{A}\mathbf{a}) + \beta(\mathbf{a}', \mathscr{A}\mathbf{b}) = (\mathbf{a}', \mathbf{v}_1) = (\mathbf{a}', \mathbf{v}) \tag{6.24a}$$

and

$$\alpha(\mathbf{b}', \mathscr{A}\mathbf{a}) + \beta(\mathbf{b}', \mathscr{A}\mathbf{b}) = (\mathbf{b}', \mathbf{v}_1) = (\mathbf{b}', \mathbf{v}). \tag{6.24b}$$

Also,

$$\mathscr{A}\mathbf{u} = \mathscr{A}\mathbf{u}_1 + \mathscr{A}\mathbf{u}'. \tag{6.25}$$

Replacing in (6.21), one gets

$$\langle B\mathbf{u}, \mathbf{v}\rangle = (\mathbf{v}_1, \mathscr{A}\mathbf{u}') = (\mathbf{v}_1, \mathscr{A}\mathbf{u}) = \alpha(\mathbf{a}, \mathbf{u}) + \beta(\mathbf{b}, \mathbf{u}), \tag{6.26}$$

because (4.44) implies $(\mathscr{A}\mathbf{a}, \mathscr{A}\mathbf{u}) = (\mathbf{a}, \mathbf{u})$ and similarly for $(\mathscr{A}\mathbf{b}, \mathscr{A}\mathbf{u})$. If $\mathbf{b}'$ is taken so that

$$\Delta = (\mathbf{a}', \mathscr{A}\mathbf{a})(\mathbf{b}', \mathscr{A}\mathbf{b}) - (\mathbf{a}', \mathscr{A}\mathbf{b})(\mathbf{b}', \mathscr{A}\mathbf{a}) = 1, \tag{6.27}$$

then

$$\begin{aligned}\langle B\mathbf{u}, \mathbf{v}\rangle = {} & (\mathbf{b}', \mathscr{A}\mathbf{b})(\mathbf{a}, \mathbf{u})(\mathbf{a}', \mathbf{v}) - (\mathbf{a}', \mathscr{A}\mathbf{b})(\mathbf{a}, \mathbf{u})(\mathbf{b}', \mathbf{v}) \\ & + (\mathbf{a}', \mathscr{A}\mathbf{a})(\mathbf{b}, \mathbf{u})(\mathbf{b}', \mathbf{v}) - (\mathbf{b}', \mathscr{A}\mathbf{a})(\mathbf{b}, \mathbf{u})(\mathbf{b}', \mathbf{v}).\end{aligned} \tag{6.28}$$

A further observation can be used to simplify expression (6.28). This is that $\mathbf{a}$ and $\mathbf{b}$ in (6.28) can be replaced by any pair of orthonormal vectors that define the same plane. If $\mathbf{a}$ is taken in the three-dimensional hyperplane for which $a_2 = 0$, then

$$\mathbf{a}(\psi, \theta) = [\cos\psi, 0, \cos\theta \sin\psi, \sin\theta \sin\psi], \tag{6.29a}$$

where $0 \leqslant \psi < \pi/2$, $0 \leqslant \theta < 2\pi$. Given $\mathbf{a}$, any vector $\mathbf{b}$ which is orthogonal simultaneously to $\mathbf{a}$ and $\mathscr{A}\mathbf{a}$ is given by

$$\begin{aligned}\mathbf{b}(\theta, \psi, \sigma) = [& -\cos\sigma \sin\psi, \sin\sigma \sin\psi, \\ & \cos\psi \cos(\theta - \sigma), \cos\psi \sin(\theta - \sigma)],\end{aligned} \tag{6.29b}$$

where σ is an additional parameter satisfying $0 \leqslant \sigma < 2\pi$. A similar argument for $\mathbf{a}'$ and $\mathbf{b}'$, yields

$$\mathbf{a}'(\theta', \psi') = \mathbf{a} \cos\psi' + \mathbf{b} \cos\theta' \sin\psi' - \mathscr{A}\mathbf{b} \sin\theta' \sin\psi' \tag{6.30a}$$

and

$$\begin{aligned}\mathbf{b}'(u, \psi', \sigma') = {} & \mathbf{a} \operatorname{ctg}\sigma' \csc\theta' \csc\psi' + \mathscr{A}\mathbf{a} \csc\psi' \csc\theta' \\ & - \mathbf{b} \operatorname{ctg}\psi' \csc\psi'(\operatorname{ctg}\sigma' \operatorname{ctg}\theta' + 1) \\ & + \mathscr{A}\mathbf{b} \operatorname{ctg}\psi' \csc\psi'(\operatorname{ctg}\sigma' - \operatorname{ctg}\theta'),\end{aligned} \tag{6.30b}$$

where θ', ψ' and σ' are additional parameters that satisfy the same restrictions as θ, ψ and σ, except that $\theta' \neq 0$, $\psi' \neq 0$ and $\sigma' \neq 0$. Note also that

$$(\mathbf{b}', \mathscr{A}\mathbf{b}) = \operatorname{ctg}\psi' \csc\psi'(\operatorname{ctg}\sigma' - \operatorname{ctg}\theta'), \tag{6.31a}$$

$$(\mathbf{a}', \mathscr{A}\mathbf{b}) = -\sin\psi' \sin\theta', \tag{6.31b}$$

$$(\mathbf{a}', \mathscr{A}\mathbf{a}) = 0; \qquad (\mathbf{b}', \mathscr{A}\mathbf{a}) = \csc\psi' \csc\theta'. \tag{6.31c}$$

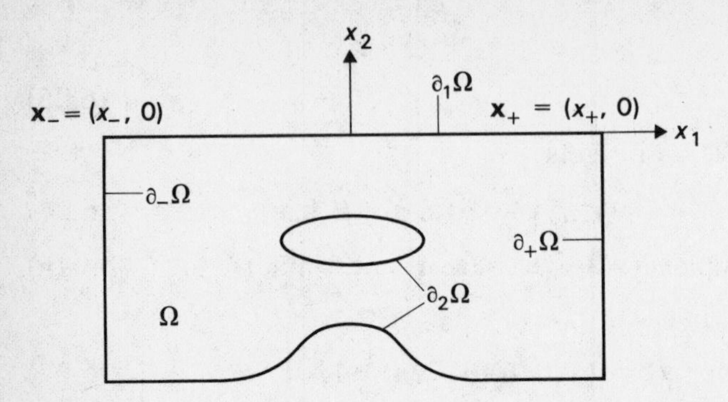

Fig. 6.1

The Green's formulas we have studied so far are pointwise in a very restricted sense. They depend on the value of the function and its normal derivatives only. In many applications they may depend on the tangential derivatives as well. An example which occurs in the linearized theory of free surface flows (see, for example, Mei and Chen [97]) is given next.

Let Ω be as in Fig. 6.1 and

$$D = \{u \in H^s(\Omega) \mid u \in H^{s+(1/2)}(\partial_1\Omega)\}, \tag{6.32}$$

where $s \geq 2$. Consider again the operator

$$\langle Pu, v\rangle = -\int_\Omega v\Delta u \, d\mathbf{x} \tag{6.33}$$

so that

$$\langle Au, v\rangle = \int_{\partial\Omega} \left\{u\frac{\partial v}{\partial n} - v\frac{\partial u}{\partial n}\right\} d\mathbf{x}. \tag{6.34}$$

Define

$$I_1 = \left\{u \in D \;\middle|\; \alpha\frac{\partial^2 u}{\partial x_1\,\partial x_1} + \frac{\partial u}{\partial x_2} = 0, \quad \text{on} \quad \partial_1\Omega \quad \text{while} \quad \frac{\partial u}{\partial n} = 0, \quad \text{on} \quad \partial_2\Omega\right\} \tag{6.35a}$$

and

$$I_2 = \{u \in D \mid u = 0, \quad \text{on} \quad \partial\Omega\}. \tag{6.35b}$$

It is not difficult to see that the pair $\{I_1, I_2\}$ is a canonical decomposition of D with respect to P. Clearly, $I_2 \subset D$ is a commutative subspace for P. Also, when $u \in I_1$ and $v \in I_2$, one has

$$\langle Au, v\rangle = \alpha \int_{\partial_1 \Omega} \left\{ v \frac{\partial^2 u}{\partial x_1\, \partial x_1} - u \frac{\partial^2 v}{\partial x_1\, \partial x_1} \right\} dx_1$$

$$= \alpha \left[v \frac{\partial u}{\partial x_1} - u \frac{\nabla v}{\partial x_1} \right]_{x_-}^{x_+} = 0. \tag{6.36}$$

Hence, I_1 is also commutative. Given any $u \in D$, let $u_1 \in I_1$ and $u_2 \in I_2$ be such that

$$u_1 = u; \quad \frac{\partial u_2}{\partial x_2} = \alpha \frac{\partial^2 u}{\partial x_1\, \partial x_1} + \frac{\partial u}{\partial x_2}, \quad \text{on} \quad \partial_1 \Omega. \tag{6.37}$$

Now the traces u and $\alpha(\partial^2 u/\partial x_1\, \partial x_1) + (\partial u/\partial x_2)$ on $\partial_1 \Omega$, span $H^{s+(1/2)}(\partial_1 \Omega)$ and $H^{s-(3/2)}(\partial_1 \Omega)$, respectively. It can be seen, using this fact, that the choice (6.37) is possible for every $u \in D$. This in turn implies that $D = I_1 + I_2$. In view of Definition 4.5, it is clear that the pair $\{I_1, I_2\}$ is a canonical decomposition of D.

The unique operator $B : D \to D^*$ decomposing A and satisfying (4.13) is given by formula (4.23):

$$\langle Bu, v\rangle = -\int_{\partial \Omega} v \frac{\partial u}{\partial n}\, d\mathbf{x} - \alpha \int_{\partial_1 \Omega} v \frac{\partial^2 u}{\partial x_1\, \partial x_1}\, d\mathbf{x}. \tag{6.38}$$

This operator is not normal because of the term

$$\int_{\partial_1 \Omega} v \left(\alpha \frac{\partial^2 u}{\partial x_1\, \partial x_1} + \frac{\partial u}{\partial x_2} \right) dx_1, \tag{6.39}$$

whose order in the tangential derivative is greater than that in the normal direction.

6.2 Parabolic equations

For a discussion of the spaces which are suitable for the formulation of this class of problems, the reader is referred to the second volume of the treatise by Lions and Magenes [82]. In this section we simply assume that the linear space of functions D is such that the operators to be considered are well defined.

Consider the cylinder $\Omega \times [0, T]$ (Fig. 6.2) and the differential operator $\mathscr{L}$ given by (3.18). Let the linear space D be made of functions defined on $\Omega \times [0, T]$. The operator $P : D \to D^*$, is

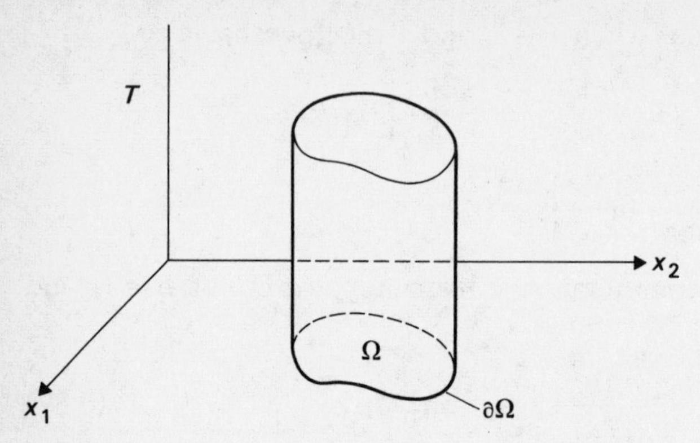

Fig. 6.2

defined by

$$\langle Pu, v\rangle = \int_{\Omega} v * \left(\frac{\partial u}{\partial t} - \mathcal{L}u\right) \mathrm{d}\mathbf{x}, \tag{6.40}$$

where the notation

$$u * v = \int_0^T u(T-t)v(t)\,\mathrm{d}t \tag{6.41}$$

is being used. Let $A = P - P^*$, then

$$A = B - B^*, \tag{6.42}$$

where

$$\langle Bu, v\rangle = \sum_{j=0}^{m-1} \int_{\partial\Omega} S_j u * C_j v\,\mathrm{d}\mathbf{x} - \int_{\Omega} v(T)u(0)\,\mathrm{d}\mathbf{x}. \tag{6.43}$$

To establish (6.42) and (6.43), equations (6.4) and (6.5) can be used. When $D = H^{r,s}(\Omega^*)$ [82], with $r \geqslant 2m$ and $r \geqslant 1$, it can be seen that $B: D \to D^*$, as given by (6.43), and $B^*: D \to D^*$ can be varied independently. Hence, (6.42) together with (6.43) yields a Green's formula for P as defined by (6.40).

The operator associated with the heat equation is obtained if $\mathcal{L} = \Delta$ is the Laplacian in (6.40). For this special case (6.43) is

$$\langle Bu, v\rangle = \int_{\partial\Omega} u * \frac{\partial v}{\partial n}\,\mathrm{d}\mathbf{x} - \int_{\Omega} v(T)u(0)\,\mathrm{d}\mathbf{x}. \tag{6.44}$$

6.3 Hyperbolic equations

The incorporation of this kind of equation in the framework here presented is similar.

Taking the region $\Omega \times [0, T]$, the differential operator $\mathscr{L}$ and the linear space of functions D, as explained before, define $P : D \to D^*$ by

$$\langle Pu, v\rangle = \int_\Omega v * \left(\frac{\partial^2 u}{\partial t^2} - \mathscr{L}u\right) \mathbf{dx} \tag{6.45}$$

with the convention (6.41). A Green's formula for this operator is obtained if $B : D \to D^*$ is given by

$$\langle Bu, v\rangle = \sum_{j=0}^{m-1} \int_{\partial\Omega} S_j u * C_j v \, \mathbf{dx} - \int_\Omega \left\{v(T)\frac{\partial u}{\partial t}(0) + \frac{\partial v}{\partial t}(T)u(0)\right\}. \tag{6.46}$$

6.4 Elasticity

Let the elastic tensor C_{ijpq} be $\mathfrak{C}^\infty(\Omega)$, satisfying the usual symmetry conditions [98]

$$C_{ijpq} = C_{pqij} = C_{jipq} \tag{6.47}$$

and being strongly elliptic, i.e.

$$C_{ijpq}\xi_i n_j \xi_p \eta_q > 0 \quad \text{whenever } \|\boldsymbol{\xi}\| \neq 0, \|\boldsymbol{\eta}\| \neq 0. \tag{6.48}$$

6.4.1 Static and periodic motions

Let $D = \mathbf{H}^s(\Omega) = H^s(\Omega) \oplus H^s(\Omega) \oplus H^s(\Omega)$, $s \geq 2$. Define

$$\tau_{ij}(\mathbf{u}) = C_{ijpq}\frac{\partial u_p}{\partial x_q}, \quad \text{on} \quad \Omega, \tag{6.49}$$

$$\mathscr{L}_i(\mathbf{u}) = \frac{\partial \tau_{ij}}{\partial x_j} + \rho\omega^2 u_i, \quad \text{on} \quad \Omega. \tag{6.50}$$

Here the density, ρ, is a function of position belonging to $\mathfrak{C}^\infty(\Omega)$, while ω is a constant. The case $\omega = 0$ is associated with elastostatics.

Let $P : D \to D^*$ be

$$\langle Pu, v\rangle = \int_\Omega v_i \mathscr{L}_i(\mathbf{u}) \, \mathbf{dx}. \tag{6.51}$$

Then, $A = P - P^*$ is given by

$$\langle Au, v\rangle = \int_{\partial\Omega} \{v_i T_i(\mathbf{u}) - u_i T_i(\mathbf{v})\} \, \mathbf{dx}, \tag{6.52}$$

where

$$T_i(\mathbf{u}) = \tau_{ij}(\mathbf{u})n_j. \tag{6.53}$$

An operator $B: D \to D^*$ that decomposes A is

$$\langle Bu, v\rangle = -\int_{\partial\Omega} u_i T_i(\mathbf{v})\, d\mathbf{x}. \tag{6.54}$$

There are many more.

6.4.2 Dynamics

Let D be a suitable linear space of functions defined on $\Omega \times [0, T]$. Define

$$\langle Pu, v\rangle = \int_{\Omega} v_i * \left(\rho \frac{\partial^2 u_i}{\partial t^2} - \mathscr{L}_i \mathbf{u}\right) d\mathbf{x}, \tag{6.55}$$

where the conventions (6.41) and (6.50) (with $\omega = 0$) are used. Then, $A = P - P^*$ is given by

$$\langle Au, v\rangle = \int_{\partial\Omega} \{u_i * T_i(v) - v_i * T_i(\mathbf{u})\}\, d\mathbf{x} + \int_{\Omega} \rho\left\{v_i(0)\frac{\partial u_i}{\partial t}(T) + \frac{\partial v_i}{\partial t}(0)u_i(T) - u_i(0)\frac{\partial v_i}{\partial t}(T) - \frac{\partial u_i}{\partial t}(0)v_i(T)\right\} d\mathbf{x}. \tag{6.56}$$

Many operators that decompose A can be constructed. One such operator is

$$\langle Bu, v\rangle = \int_{\partial\Omega} u_i * T_i(\mathbf{v})\, d\mathbf{x} - \int_{\Omega} \left\{u_i(0)\frac{\partial v_i}{\partial t}(T) + \frac{\partial u_i}{\partial t}(0)v_i(T)\right\} d\mathbf{x}. \tag{6.57}$$

6.5 Stokes' problem

The basic equations for this problem are [99, 100]:

$$-\Delta u + \nabla p = 0 \quad \text{in} \quad \Omega, \tag{6.58a}$$

$$\nabla \cdot \mathbf{u} = 0 \quad \text{in} \quad \Omega. \tag{6.58b}$$

Function spaces suitable for the formulation of these equations were discussed previously [39]. However, the formulation can be simplified.

Let

$$D = \{\{\mathbf{u}, p\} \mid \mathbf{u} \in \mathbf{H}^s(\Omega) \quad \& \quad p \in H^s(\Omega)\}, \tag{6.59}$$

where $s \geq 2$. Elements of this space are pairs $\{\mathbf{u}, p\}$, where $\mathbf{u}$ is a vector valued function while p is scalar valued. The notation $\hat{u}$ will be used for such pairs; thus, $\hat{u} = \{\mathbf{u}, p\}$. Let $\hat{v} = \{\mathbf{v}, q\}$ be another arbitrary element of D and define $P : D \to D^*$ by

$$\langle P\hat{u}, \hat{v}\rangle = \int_\Omega \{\mathbf{v} \cdot (\nu \Delta \mathbf{u} - \nabla p) + q \nabla \cdot \mathbf{u}\} \, d\mathbf{x}. \tag{6.60}$$

Then

$$\langle A\hat{u}, \hat{v}\rangle = \int_{\partial\Omega} \left\{\mathbf{u} \cdot \left(q\mathbf{n} - \nu \frac{\partial \mathbf{v}}{\partial n}\right) - \mathbf{v} \cdot \left(p\mathbf{n} - \nu \frac{\partial \mathbf{u}}{\partial n}\right)\right\} d\mathbf{x}. \tag{6.61}$$

A Green's formula which is suitable for the case when $\mathbf{u}$ is prescribed on $\partial\Omega$ is associated with $B : D \to D^*$ when this latter operator is defined by

$$\langle B\hat{u}, \hat{v}\rangle = \int_{\partial\Omega} \mathbf{u} \cdot \left(q\mathbf{n} - \nu \frac{\partial \mathbf{v}}{\partial n}\right) d\mathbf{x}. \tag{6.62}$$

It is easy to see that B decomposes A.

7 Illustrations of jump operators

For each of the examples presented in Chapter 6 one can construct jump operators and corresponding Green's formulas for the case when discontinuous fields are admitted and jump conditions are prescribed. The application of equations (5.37) or, equivalently, equations (5.38) to each of those cases is a simple exercise.

The examples presented in Chapter 6 refer to problems formulated in continuous fields. Thus, the functions belonging to the linear space D are continuous. When applying the theory of Chapter 5 to problems formulated in discontinuous fields, it is frequently convenient to take the elements of D_R and D_E as the restrictions to R and E, respectively, of functions belonging to D.

7.1 Elliptic equations

If D is the linear space of functions used in Section 6.1, D_R and D_E can be taken as just explained. The regions R, E and Ω are illustrated in Fig. 5.1 (p. 40). Thus, the elements of D_R are the restrictions to R of functions belonging to $H^s(\Omega)$, $s > 2m$. Similarly for D_E. This defines $\hat{D} = D_R \oplus D_E$. Let the operator $P_R : D_R \rightarrow D_R^*$ be

$$\langle P_R u_R, v_R \rangle = \int_R v \mathscr{L} u \, d\mathbf{x} - \sum_{j=0}^{m-1} \int_{\partial'' R} S_j u C_j v \, d\mathbf{x}; \tag{7.1}$$

$P_E : D_E \rightarrow D_E^*$ is obtained replacing R by E. Here, $\mathscr{L}u$ is given by (3.18), while $S_j u$ and $C_j v$ $(j = 0, \ldots, m-1)$ are taken as in Section 6.1. Then $\hat{P} : \hat{D} \rightarrow \hat{D}^*$ is given by equation (5.1).

Using the results on Green's formulas of Section 6.1 and applying equation (5.7), one gets

$$\langle \hat{A}\hat{u}, \hat{v} \rangle = \sum_{j=0}^{m-1} \left\{ \int_{\partial' R} (S_j u C_j v - S_j v C_j u) \, d\mathbf{x} + \int_{\partial' E} (S_j u C_j v - S_j v C_j u) \, d\mathbf{x} \right\}, \tag{7.2}$$

using the conventions introduced in Chapter 5 regarding the components of the functions and the sense of the normal vectors. Using the fact that the C_js and the S_js can be varied independently, one gets that

$$\hat{N}_A = \{\hat{u} \in \hat{D} \mid C_i u_R = C_i u_E = S_j u_R = S_j u_E = 0,$$
$$j = 0, 1, \ldots, m-1, \quad \text{on} \quad \partial' R = \partial' E\}. \quad (7.3)$$

In addition, the fact that $\{C_0, \ldots, C_{m-1}, S_0, \ldots, S_{m-1}\}$ is a Dirichlet system or order $2m$ implies that

$$\hat{N}_A = \left\{\hat{u} \in \hat{D} \,\middle|\, \frac{\partial^\nu u_R}{\partial n_R^\nu} = \frac{\partial^\nu u_E}{\partial n_R^\nu} = 0; \qquad \nu = 0, 1, \ldots, 2m-1, \right.$$
$$\left. \partial' R = \partial' E \right\}. \quad (7.4)$$

The convention of writing the derivative of order zero for the function itself is adopted in (7.4).

There are many possible choices for the definition of the smoothness condition $\hat{S}$. Its adequacy depends on the particular problem considered. For illustration, here only the case when

$$\hat{S} = \left\{\hat{u} \in D \,\middle|\, \frac{\partial^\nu u_R}{\partial n_R^\nu} = \frac{\partial^\nu u_E}{\partial n_R^\nu}, \quad \nu = 0, 1, \ldots, 2m-1, \right.$$
$$\left. \text{on} \quad \partial' R = \partial' E \right\} \quad (7.5)$$

will be considered. Clearly, $\hat{S} \supset N_A$. It is more difficult to see that $\hat{S}$ is isotropic, i.e., that

$$\langle \hat{A}\hat{u}, \hat{v} \rangle = 0 \quad \forall\, \hat{u} \in \hat{S} \quad \& \quad \hat{v} \in \hat{S}. \quad (7.6)$$

First notice that equation (7.5) is equivalent to

$$\hat{S} = \{\hat{u} \in \hat{D} \mid C_j u_R = C_j u_E,\ S_j u_R = S_j u_E,$$
$$j = 0, \ldots, m-1, \quad \text{on} \quad \partial' R = \partial' E\}, \quad (7.7)$$

because the definition given in Section 6.1 involves derivatives only up to order $2m-1$. The space $H^s(\Omega)$ can be imbedded in $\hat{D}$, if the former space is identified with linear subspace $\hat{H}^s(\Omega) \subset \hat{D}$ whose elements are pairs $\{u_R, u_E\}$ with the property that u_R and u_E are restrictions, to R and E, of the same function $u \in H^s(\Omega)$. Clearly, $\hat{H}^s(\Omega) \subset \hat{S}$. Let $\hat{u} = \{u_R, u_E\} \in \hat{S}$ and $\hat{v} = \{v_R, v_E\} \in \hat{S}$ be otherwise arbitrary. Take $\hat{u}' = \{u'_R, u'_E\} \in \hat{H}^s(\Omega)$ and $\hat{v}' = \{v'_R, u'_E\} \in \hat{H}^s(\Omega)$, so that

$$u_R = u'_R \quad \text{and} \quad v_R = v'_R. \quad (7.8)$$

This choice is possible by the manner in which D_R was defined. Taking into account (7.7), it is seen that

$$C_j u'_E = C_j u'_R = C_j u_R = C_j u_E, \quad j = 0, 1, \ldots, m-1 \tag{7.9a}$$

and

$$S_j u'_E = S_j u'_R = S_j u_R = S_j u_E, \quad j = 0, 1, \ldots, m-1. \tag{7.9b}$$

Comparing with (7.2), it is seen that

$$\langle \hat{A}\hat{u}, \hat{v} \rangle = \langle \hat{A}\hat{u}', \hat{v}' \rangle. \tag{7.10}$$

However,

$$\langle \hat{A}\hat{u}', \hat{v}' \rangle = \int_\Omega \{v\mathcal{L}u - u\mathcal{L}v\} \, d\mathbf{x} - \sum_{j=0}^{m-1} \int_{\partial\Omega} \{S_j u C_j v - S_j v C_j u\} \, d\mathbf{x} = 0. \tag{7.11}$$

This shows that $\hat{S} \subset \hat{D}$ is isotropic and, hence, regular.

Applying equations (5.37) and (7.2), one gets

$$\langle \hat{J}\hat{u}, \hat{v} \rangle = \sum_{j=0}^{m-1} \int_{\partial' R} (S_j \dot{v} C_j [u] - S_j [u] C_j \dot{v}) \, d\mathbf{x}, \tag{7.12}$$

where

$$[\hat{u}] = u_E - u_R \quad \text{and} \quad \dot{v} = \tfrac{1}{2}(v_E + v_R). \tag{7.13}$$

Thus, the general Green's formula for elliptic equations acting on discontinuous fields is

$$\begin{aligned} \int_\Omega \{v\mathcal{L}u - u\mathcal{L}v\} \, d\mathbf{x} = &\sum_{j=1}^{m-1} \int_{\partial\Omega} \{S_j u C_j v - S_j v C_j u\} \, d\mathbf{x} \\ &+ \sum_{j=0}^{m-1} \int_{\partial' R} \{S_j \dot{v} C_j [u] - S_j [u] C_j \dot{v}\} \, d\mathbf{x} \\ &- \sum_{j=0}^{m-1} \int_{\partial' R} \{S_j \dot{u} C_j [v] - S_j [v] C_j \dot{u}\} \, d\mathbf{x}. \end{aligned} \tag{7.14}$$

The operator $\hat{J}$ that decomposes $\hat{A}$ is given by (7.12). Finally, recall that the normal vector on $\partial' R$ is taken pointing outwards from R.

In applications, it is frequently necessary to treat cases in which the coefficients are discontinuous across the common boundary $\partial' R = \partial' E$. A problem of this kind relevant in applications to flow through porous media was presented in Example 5.6. Applications to elastic diffraction are given in [40], for which the regions considered may be as those illustrated in Figs 7.1 and 7.2.

Fig. 7.1

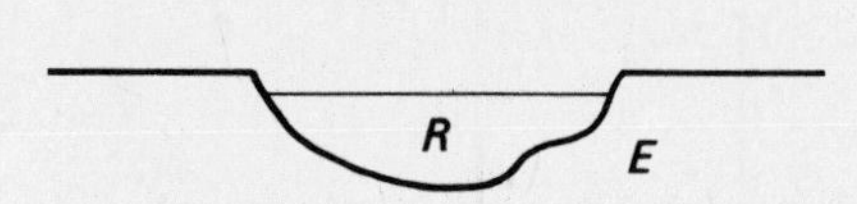

Fig. 7.2

7.2 Parabolic equations

In this case the regions (Fig. 7.3) are the neighbouring cylinders $R = R_x X[0, T]$ and $E_x X[0, T]$. Using the results of Section 7.1, the developments can be easily applied to the operator

$$Lu = \mathscr{L}u - \frac{\partial u}{\partial t}, \tag{7.15}$$

where $\mathscr{L}$ is the same as in that section. However, since the basic idea is essentially the same, it is clearer, because it is simpler, to restrict attention to the heat equation. Thus, let

$$Lu = \Delta u - \frac{\partial u}{\partial t}. \tag{7.16}$$

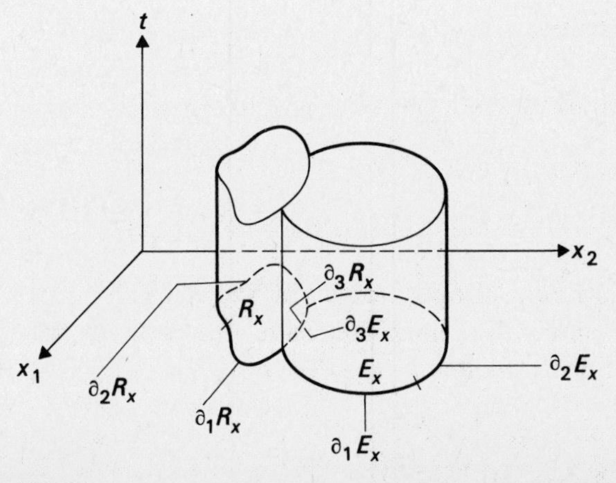

Fig. 7.3

The choice of $P_R : D_R \to D_R^*$ depends on the boundary conditions to be treated.

$$\langle P_R u_R, v_R \rangle = \int_{R_x} v * Lu \, d\mathbf{x} + \int_{\partial_1 R_x} u * \frac{\partial v}{\partial n} \, d\mathbf{x} - \int_{\partial_2 R_x} v * \frac{\partial u}{\partial n} \, d\mathbf{x} - \int_{R_x} u(0) v(T) \, d\mathbf{x}. \tag{7.17}$$

This is suitable when u and $\partial u / \partial n$ are prescribed on $\partial_1 R_x$ and $\partial_2 R_x$, respectively.

Here, as in what follows, the notation

$$u * v = \int_0^R u(T-t) v(t) \, dt \tag{7.18}$$

is adopted. $P_E : D_E \to D_E^*$ is obtained replacing R by E in (7.17).

$$\langle A_R u_R, v_R \rangle = \int_{\partial_3 R_x} \left\{ u * \frac{\partial v}{\partial n} - v * \frac{\partial u}{\partial n} \right\} d\mathbf{x}. \tag{7.19}$$

The smoothness condition can be taken as

$$\hat{S} = \{ \hat{u} \in \hat{D} \mid u_R = u_E, \quad \partial u_R / \partial n = \partial u_E / \partial n, \quad \text{on} \quad \partial_3 R \}, \tag{7.20}$$

where the subsets $\partial_i R = [0, T] \times \partial_i R_x$ $(i = 1, 2, 3)$. It is easy to see that $\hat{S}$ is regular for $P : D \to D^*$. Equations (5.37) and (7.19) yield

$$\langle \hat{J} \hat{u}, \hat{v} \rangle = \int_{\partial_3 R_x} \left\{ \frac{\partial \dot{v}}{\partial n} * [\hat{u}] - \dot{v} * \left[\frac{\partial \hat{u}}{\partial n} \right] \right\} d\mathbf{x}. \tag{7.21}$$

7.3 Hyperbolic equations

This is similar to the parabolic case. It is easy to do the general case,

$$\mathscr{L} u - \frac{\partial^2 u}{\partial t^2} = 0, \tag{7.22}$$

where $\mathscr{L}$ is the same differential operator as in Section 7.1.

However, again for simplicity, only the wave equation is presented. The results are listed below.

$$\left. \begin{aligned} &\text{(a)} \quad R = R_x \times [0, T]; \quad E = E_x \times [0, T]. \\ &\text{(b)} \quad Lu = \nabla^2 u - (\partial^2 u / \partial t^2). \end{aligned} \right\} \tag{7.23}$$

(c) $P_R : D_R \to D_R^*$ is

$$\langle P_R u_R, v_R \rangle = \int_{R_x} vLu \, \mathrm{d}\mathbf{x} + \int_{\partial_1 R_x} u * \frac{\partial v}{\partial n} \mathrm{d}\mathbf{x} - \int_{\partial_2 R_x} v * \frac{\partial u}{\partial n} \mathrm{d}\mathbf{x}$$
$$- \int_{R_x} \{u(0)v'(T) + u'(0)v(T)\} \, \mathrm{d}\mathbf{x}, \qquad (7.24)$$

where the primes stand for the partial derivatives with respect to t. To obtain $P_E : D_E^* \to D_E$, R has to be replaced by E in (7.16).
(d) The operator $\hat{J} : \hat{D} \to \hat{D}^*$ that decomposes $\hat{A}$ is again given by (7.21).

7.4 Elasticity

In order to formulate problems of elasticity, the elastic tensor C_{ijpq} is assumed to be defined in regions R and E, and to be $\mathfrak{C}^\infty(R)$ and $\mathfrak{C}^\infty(E)$, separately. In addition, the usual symmetry conditions (6.47) and strong ellipticity (6.48) will be taken for granted.

7.4.1 Static and periodic motions

Elements of the linear spaces D_R and D_E, will be restrictions to R and E of vector-valued functions belonging to $H^s(\Omega)$, respectively. Here, $s \geq 2$. Using the notation introduced in Section 6.4:

(i) $P_R : D_R \to D_R^*$ is

$$\langle P_R u_R, v_R \rangle = \int_R \mathbf{v} \cdot \mathbf{L}(\mathbf{u}) \, \mathrm{d}\mathbf{x} + \int_{\partial_1 R} \mathbf{u} \cdot \mathbf{T}(\mathbf{v}) \, \mathrm{d}\mathbf{x} - \int_{\partial_2 R} \mathbf{v} \cdot \mathbf{T}(\mathbf{u}) \, \mathrm{d}\mathbf{x} \qquad (7.25)$$

and $P_E : D_E \to D_E^*$ is obtained replacing R by E in (7.25).
(ii) $A_R : D_R \to D_R^*$ is

$$\langle A_R u_R, v_R \rangle = \int_{\partial_3 R} \{\mathbf{v} \cdot \mathbf{T}(\mathbf{u}) - \mathbf{u} \cdot \mathbf{T}(\mathbf{v})\} \, \mathrm{d}\mathbf{x}. \qquad (7.26)$$

(iii) $$\hat{N}_A = \{\hat{u} \in \hat{D} \mid \mathbf{u}_R = \mathbf{u}_E = \mathbf{T}_R(\mathbf{u}_R) = \mathbf{T}_E(\mathbf{u}_E) = 0, \quad \text{on} \quad \partial_3 R = \partial_3 E\}. \qquad (7.27)$$

It must be observed that the traction $\mathbf{T}_R(\mathbf{u}_R)$ on $\partial_3 R$ must be computed using not only the unit normal vector pointing outwards from R, but also the limiting values of the elastic tensor associated with that side, and similarly for $\mathbf{T}_E(\mathbf{u}_E)$. Also, to establish (7.27) one has to use the fact that $\mathbf{u}$ and $\mathbf{T}(\mathbf{u})$ can be varied independently on the boundary: a fact that is granted by the assumed strong ellipticity of the elastic tensor.

(iv) The usual smoothness relation is

$$\hat{S}=\{\hat{u}\in\hat{D}\mid \mathbf{u}_R=\mathbf{u}_E \quad \text{and} \quad \mathbf{T}_R(\mathbf{u}_R)+\mathbf{T}_E(\mathbf{u}_E)=0, \quad \text{on} \quad \partial_3 R=\partial_3 E\}. \tag{7.28}$$

It must be pointed out that $\hat{S}$ is the set of displacement fields which satisfy the usual continuity conditions for displacements and tractions across $\partial_3 R=\partial_3 E$. The continuity of tractions in our notations is $\mathbf{T}_R(\mathbf{u}_R)+\mathbf{T}_E(\mathbf{u}_E)=0$, because the unit normal vector used for computing $\mathbf{T}_R$ is opposite to that used in $\mathbf{T}_E$.

It is easy to see that $\hat{S}$ is regular. Application of equation (5.37) yields

$$\langle \hat{J}\hat{u}, \hat{v}\rangle=\int_{\partial_3 R}\{[\mathbf{u}]\cdot(\mathbf{T}(\mathbf{v}))_a-\dot{\mathbf{v}}\cdot[\mathbf{T}(\mathbf{u})]\}\,\mathrm{d}\mathbf{x}. \tag{7.29}$$

Here

$$\dot{\mathbf{v}}=\mathbf{v}_E+\mathbf{v}_R; \qquad (\mathbf{T}(\mathbf{v}))_a=(\mathbf{T}_R(\mathbf{u}_R)-\mathbf{T}_E(\mathbf{u}_E))/2 \tag{7.30a}$$

and

$$[\mathbf{u}]=\mathbf{u}_E-\mathbf{u}_R; \qquad [\mathbf{T}(\mathbf{u})]=-\mathbf{T}_E(\mathbf{u}_E)-\mathbf{T}_R(\mathbf{u}_R). \tag{7.30b}$$

Observe that

$$\mathbf{T}_E(\mathbf{u}_E)=C^E_{ijpq}\frac{\partial u_{Ep}}{\partial x_q}n_{Ej}=-C^E_{ijpq}\frac{\partial u_{Ep}}{\partial x_q}n_{Rj}. \tag{7.31}$$

Thus $(\mathbf{T}(\mathbf{v}))_a$ and $[\mathbf{T}(\mathbf{u})]$ represent the usual average and jump of the tractions when they are computed using the unit normal vector associated with only one of the sides.

PART 2

Boundary methods

8 Scope

To fix ideas we first consider a simple example. Take Laplace or Poisson's equation in a bounded region Ω, illustrated in Fig. 8.1, and subjected to boundary conditions of Dirichlet type:

$$\Delta u = f_\Omega, \quad \text{in} \quad \Omega \tag{8.1a}$$

and

$$u = f_{\partial\Omega}, \quad \text{on} \quad \partial\Omega, \tag{8.1b}$$

where f_Ω and $f_{\partial\Omega}$ are given functions. For the time being, attention will be restricted to the case of real coefficients.

In general, the application of boundary methods requires the transformation of equation (8.1a) into a homogeneous equation. This can be achieved by introducing a particular solution U of equation (8.1a). Thus

$$\Delta U = f_\Omega, \quad \text{in} \quad \Omega. \tag{8.2a}$$

In applications the construction of such a function U is not difficult, because boundary conditions are not prescribed. In particular, when a fundamental solution is available, U can be obtained by quadrature.

In addition, let V be a function such that

$$V = f_{\partial\Omega}, \quad \text{on} \quad \partial\Omega. \tag{8.2b}$$

Then, the Dirichlet problem (8.1) is equivalent to

$$\Delta(u - U) = 0, \quad \text{in} \quad \Omega \tag{8.3a}$$

and

$$u = V, \quad \text{on} \quad \partial\Omega. \tag{8.3b}$$

In order to formulate this problem precisely, it is necessary to define a space D of admissible functions. Consider Sobolev space $H^s(\Omega)$, where s is any real number ($-\infty < s < \infty$). It is well known that the trace operator (i.e. the boundary values) is not defined for

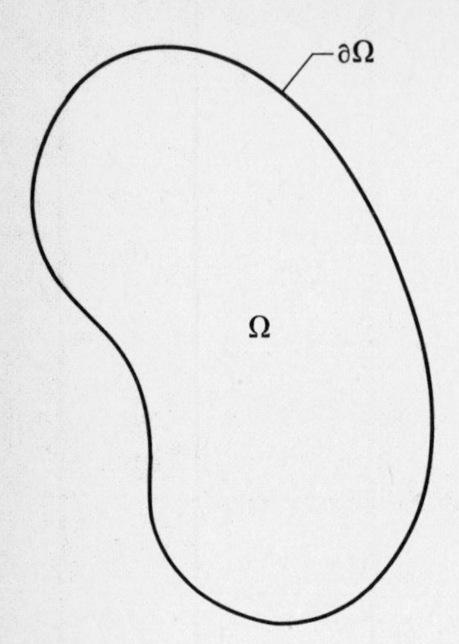

Fig. 8.1

some elements of $H^s(\Omega)$ when $s \leq \frac{1}{2}$ [82, 83]. However, there is a wide class of functions of $H^s(\Omega)$ for which this trace is defined and belongs to $H^{s-(1/2)}(\partial\Omega)$. Thus, define

$$D^e = \bigcup_s H^s(\Omega) \tag{8.4}$$

and

$$D = \{u \in D^e \mid \gamma_0 u \in \bigcup_s H^s(\partial\Omega)\}, \tag{8.5}$$

where $\bigcup$ is the union symbol used in set theory and γ_0 stands for the trace of u on $\partial\Omega$. Frequently, for simplicity the symbol γ_0 will be omitted when it is clear from the context that we refer to the boundary values. It can be noticed that the linear space D defined by (8.5) is not closed. Indeed, a metric is not defined in the whole space.

Let

$$N_P = \{u \in D \mid \Delta u = 0 \quad \text{in} \quad \Omega\} \tag{8.6}$$

and

$$I = \{u \in D \mid u = 0, \quad \text{on} \quad \partial\Omega\}. \tag{8.7}$$

In order for definition (8.6) to make sense, the Laplacian must be interpreted in the sense of distributions [82, 83].

Then, Dirichlet problem can be formulated as a problem with linear restrictions. Given any $U \in D$ and $V \in D$ (these functions can be taken as data of the problem), find an element $u \in D$ such that

$$u - U \in N_P \quad \text{and} \quad u - V \in I. \tag{8.8}$$

The first of equations (8.8) is equivalent to (8.3a) and the second, to (8.3b).

A first advantage of formulating the problem in this manner is connected with its existence properties. Clearly equation (8.3b) is equivalent to $u - U = V - U$ on $\partial\Omega$. By well-known results on the existence of solution [82], this problem possesses a unique solution. Indeed, given $U \in D$ and $V \in D$ there are real numbers r and s such that $U \in H^r(\Omega)$ and the trace $\gamma_0(V - U) \in H^s(\partial\Omega)$. Then, $u - U \in H^{s+(1/2)}(\Omega)$. Therefore, $u = U + (u - U)$ belongs to $H^t(\Omega)$ where $t = \min\{r, s + \frac{1}{2}\}$. This shows $u \in D$.

The above discussion also shows that there is no lack of generality by restricting attention to the homogeneous case, i.e.,

$$\Delta u = 0, \quad \text{in} \quad \Omega \tag{8.9a}$$

and

$$u = f_{\partial\Omega}, \quad \text{on} \quad \partial\Omega. \tag{8.9b}$$

The boundary method to be applied is based on the continuous dependence of solutions on their boundary values. In principle it can be applied when the space of admissible functions D is given by (8.5). However, this would lead to the consideration of inner products in the space of boundary values $H^s(\partial\Omega)$ with arbitrary s, which may be inconvenient in numerical applications. It is preferable to keep the computations in $\mathscr{L}^2(\partial\Omega) = H^\circ(\partial\Omega)$, which, as will be seen, leads to least-squares fitting. This, can be achieved if attention is restricted to functions with boundary values belonging to $H^\circ(\partial\Omega) = \mathscr{L}^2(\partial\Omega)$. When this condition is incorporated in the definition of the space of admissible functions, one gets

$$D = \{u \in D^e \mid \gamma_0 u \in H^\circ(\partial\Omega)\}. \tag{8.10a}$$

This is again a linear space which is not closed.

In addition, in many applications it is necessary to compute the normal derivative $\partial u/\partial n$ on the boundary $\partial\Omega$. Similar considerations lead to the requirement that $\partial u/\partial n$ belong to $H^\circ(\partial\Omega) = \mathscr{L}^2(\partial\Omega)$. When these two requirements are incorporated in the definition of the space of admissible functions, equation (8.5) becomes

$$D=\{u\in D^e \mid \gamma_0 u\in H^\circ(\partial\Omega),\ \gamma_1 u\in H^\circ(\partial\Omega)\}. \tag{8.11a}$$

here, as is customary, $\gamma_1 u$ stands for the trace of the normal derivative on $\partial\Omega$. This is again a linear space.

General results on the existence and continuity properties of solutions of elliptic equations [82] imply that any harmonic function u whose trace $\gamma_0 u$ belongs to $H^\circ(\partial\Omega)$ is necessarily a member of $H^{1/2}(\Omega)$. Therefore $N_P\subset H^{1/2}(\Omega)$ in this case. Moreover, due to the continuity properties just mentioned, N_P is a closed subspace of $H^{1/2}(\Omega)$. This will be represented by $N^{1/2}(\Omega)$. Thus

$$N_P=N^{1/2}(\Omega) \tag{8.10b}$$

when D is defined by equation (8.10a). Similarly, when equation (8.11a) holds, corresponding properties imply that

$$N_P=N^{3/2}(\Omega), \tag{8.11b}$$

where $N^{3/2}(\Omega)$ is the subspace of harmonic functions belonging to $H^{3/2}(\Omega)$ which can also be shown to be closed.

If $\mathfrak{B}=\{w_1, w_2, \ldots\}\subset N^{3/2}$ is a system of harmonic functions in Ω which spans $N^{3/2}$ then it also spans $N^{1/2}$, because $N^{3/2}\subset N^{1/2}$ is dense in $N^{1/2}$. Thus, let $\mathfrak{B}=\{w_1, w_2, \ldots\}\subset N^{3/2}$ be such a system. Then, given $u\in N^{1/2}$, there is a sequence of approximations

$$u^N=\sum_{n=1}^{N} a_n^N w_n, \quad N=1,2,\ldots, \tag{8.12}$$

such that

$$u^N\to u \quad \text{in} \quad H^{1/2}(\Omega). \tag{8.13}$$

Notice that a_n^N depends on the number of terms, N, of the approximation. This is essential in order for (8.13) to hold. When a_n^N is independent of n, (8.12) becomes a series and the approximation by a series cannot be granted. A sufficient condition is that the system of functions $\mathfrak{B}$ be orthonormal or, more generally, Hilbertian–Besselian [6]. This fact explains some of the difficulties that were encountered in applications to electromagnetic field studies [78].

In order for representation (8.12) to be useful, it is necessary to have a procedure for deriving the coefficients a_n^N from boundary data only. This is indeed possible. General results on the existence and continuity properties of solutions of elliptic equations [82] show that the range of the traces $\gamma_0 u$ when $u \in N^{1/2}(\Omega)$ is $H^\circ(\partial\Omega)$. Given $u \in N^{1/2}(\Omega)$, if the coefficients a_n^N are chosen so that

$$\gamma_0 u^N \to \gamma_0 u \quad \text{on} \quad H^\circ(\partial\Omega) = \mathscr{L}^2(\partial\Omega), \tag{8.14}$$

then (8.13) holds necessarily.

If $\mathfrak{B} = \{w_1, w_2, \ldots\} \subset N^{3/2}(\Omega)$ spans $N^{3/2}(\Omega)$, then the continuity properties of elliptic equations imply that the associated system of traces $\{\gamma_0 w_1, \gamma_0 w_2, \ldots\}$ spans $H^1(\partial\Omega)$, which in turn implies that

$$\{\gamma_0 w_1, \gamma_0 w_2, \ldots\} \quad \text{spans} \quad H^\circ(\partial\Omega). \tag{8.15}$$

These remarks show that the coefficients a_n^N can be chosen so that (8.14) holds. Clearly, to this end it will be sufficient to take $\gamma_0 u^N$ as the projection (in the $H^\circ(\partial\Omega)$ sense) of the boundary values $\gamma_0 u \in H^\circ(\partial\Omega)$ on the subspace spanned by $\{w_1, \ldots, w_N\}$. Therefore, the coefficients can be computed by standard procedures for projecting on a subspace. This yields a least-squares method because only $\mathscr{L}^2(\partial\Omega) = H^\circ(\partial\Omega)$ inner products are being used.

This procedure leads to the system of equations:

$$\sum_{n=1}^{N} M_{nm} a_n^N = c_m, \quad m = 1, \ldots, N \tag{8.16}$$

for the coefficients a_n^N occurring in equation (8.12). Here

$$M_{nm} = \int_{\partial\Omega} w_n w_m \, \mathrm{d}\mathbf{x} \tag{8.17a}$$

and

$$c_m = \int_{\partial\Omega} f_{\partial\Omega} w_m \, \mathrm{d}\mathbf{x}. \tag{8.17b}$$

In order for this system to be invertible it is required that the system of traces $\{w_1, \ldots, w_N\} \subset H^\circ(\partial\Omega)$ be linearly independent. Then, the matrix M_{nm} can be seen to be Hermitian. Clearly, if the problem possesses a solution this procedure is convergent, since the fact that $\gamma_0 u^N$ is the projection of $\gamma_0 u$ on the subspace spanned by $\{w_1, \ldots, w_N\}$ grants that (8.14) is satisfied. However, numerical difficulties may still occur (see, for example, Aleksidze [101]). Such difficulties can be avoided altogether if the system is Hilbertian–Besselian from the start.

Computation of the boundary values may need a special device. If the normal derivatives $\partial u/\partial n$ are required on the boundary, in the case of Dirichlet problem, they cannot be obtained directly from the approximating sequence u^N. Indeed, as mentioned, from $u^N \to f_{\partial\Omega}$ on $H^\circ(\partial\Omega)$, one can only grant that $u^N \to u$ in $H^{1/2}(\Omega)$. Hence $(\partial u^N/\partial n) \to (\partial u/\partial n)$ in the metric of $H^{-1}(\partial\Omega)$ [82]; thus, $\partial u^N/\partial n$ diverges in $H^\circ(\partial\Omega)$ in general. As a matter of fact, $(\partial u/\partial n) \in H^{-1}(\partial\Omega)$ is only defined as a continuous functional on $H^1(\partial\Omega)$. The procedure to be presented can be extended to supply a method for computing this functional, but for simplicity such extension is not included here.

This can be avoided if the space of admissible functions D is restricted. Let it be defined by equations (8.11). In this case the boundary data are more smooth, necessarily; indeed, $f_{\partial\Omega} \in H^1(\partial\Omega)$ and $(\partial u/\partial n) \in H^\circ(\partial\Omega)$. Define the bilinear functional

$$\langle Au, v\rangle = \int_{\partial\Omega} \left\{ v\frac{\partial u}{\partial n} - u\frac{\partial v}{\partial n} \right\} d\mathbf{x}. \tag{8.18}$$

This is well defined for every $u \in D$ and $v \in D$, because their traces belong to $H^\circ(\partial\Omega)$. Observe that

$$\langle Au, v\rangle = (v, \partial u/\partial n)_\partial - (u, \partial v/\partial n)_\partial, \tag{8.19}$$

where we have written $(\ ,\ )_\partial$ for the inner product in $H^\circ(\partial\Omega)$. Also $\langle Au, v\rangle = 0$ whenever $u \in N_P$ and $v \in N_P$. This yields the reciprocity relation

$$(v, \partial u/\partial n)_\partial = (u, \partial v/\partial n)_\partial. \tag{8.20}$$

Using it, the following approximating sequence will be constructed:

$$\sum_{n=1}^{N} b_n^N w_n \to \frac{\partial u}{\partial n} \quad \text{in} \quad H^\circ(\partial\Omega). \tag{8.21}$$

When $u \in N^{3/2}(\Omega)$ the trace $\gamma_1 u$ (i.e. the normal derivative on the boundary) spans

$$\{1\}^\perp \subset \mathrm{H}^\circ(\partial\Omega) = \mathfrak{L}^2(\partial\Omega). \tag{8.22}$$

Here, $\{1\}^\perp$ stands for the orthogonal complement in $H^\circ(\partial\Omega)$ of the constant function. Thus, $(\partial u/\partial n) \in \{1\}^\perp$ if and only if $(\partial u/\partial n) \in H^\circ(\partial\Omega)$ and

$$\int_{\partial\Omega} \frac{\partial u}{\partial n}\, d\mathbf{x} = 0. \tag{8.23}$$

Taking $\mathfrak{B} = \{w_1, w_2, \ldots\} \subset N^{3/2}(\Omega)$ as before, and in view of (8.15), given any $u \in N^{3/2}(\Omega)$, it is possible to construct an approximation satisfying (8.21). Again, for such representation to be useful it will be required to have a procedure for computing b_n^N using boundary data only. This will be based on the use of the reciprocity relation (8.20). It is interesting to observe a general feature of representation (8.21): namely, that the normal derivative $\partial u/\partial n$ of the sought solution is not approximated by the normal derivatives of the basic system $\mathfrak{B}$, but by its boundary values, instead. The coefficients b_n^N will be chosen so that

$$\left\| \frac{\partial u}{\partial n} - \sum_{n=1}^{N} b_n^N w_n \right\|^2 \tag{8.24}$$

is minimized.

This leads us to take the projection of $\partial u/\partial n$ on the space spanned by $\{w_1, \ldots, w_N\} \subset H^\circ(\partial\Omega)$. This requires the orthogonality condition

$$\left(\frac{\partial u}{\partial n} - \sum_{n=1}^{N} b_n^N w_n, w_m \right) = 0, \quad m = 1, \ldots, N \tag{8.25}$$

to be satisfied. Expanding (8.25), one gets

$$\sum_{n=1}^{N} K_{nm} b_n^N = d_m, \tag{8.26}$$

where

$$K_{nm} = \int_{\partial\Omega} w_n w_m \, d\mathbf{x}, \quad n, m = 1, \ldots, N \tag{8.27a}$$

and

$$d_m = \int_{\partial\Omega} \frac{\partial u}{\partial n} w_m \, d\mathbf{x} = \int_{\partial\Omega} f_{\partial\Omega} \frac{\partial w_m}{\partial n} \, d\mathbf{x}, \quad m = 1, \ldots, N. \tag{8.27b}$$

Observe that the use of the reciprocity relation (8.20) has permitted us to express d_m in terms of boundary data only.

An additional point must be mentioned. In order for the approximating sequence $\sum_{n=1}^{N} b_n^N w_n$ to be convergent, it is necessary that the solution $(\partial u/\partial n) \in H^\circ(\partial\Omega)$. This is granted if $f_{\partial\Omega} \in \mathrm{H}^1(\partial\Omega)$. Alternatively, this condition can be expressed in matrix form. Let $\boldsymbol{K}^N$ be the $N \times N$ square matrix whose elements are given by (8.27a). Similarly $\mathbf{d}^N$ is the $1 \times N$ vector defined by (8.27b). Assume that the system of traces $\{w_1, \ldots, w_N\} \subset H^1(\partial\Omega)$ is linearly independent, which is required in order for the system (8.26) to be invertible, and denote by $(\boldsymbol{K}^N)^{-1}$ the inverse of $\boldsymbol{K}^N$. Then, the sequence of real numbers

$$\left\|\sum_{n=1}^{N} b_n^N w_n\right\|^2 = \mathbf{d}^N \cdot (\mathbf{K}^N)^{-1}, \quad \mathbf{d}^N \geqslant 0, \quad N = 1, 2, \ldots \tag{8.28}$$

is non-negative and increasing. Convergence of the approximating sequence is granted when the sequence (8.28) is bounded. The meaning of this condition is more easily understood by observing that when the system of traces $\{w_1, w_2, \ldots\}$ is orthonormal (i.e., $K_{nm} = \delta_{nm}$), in which case the coefficients d_n are independent of N, it becomes

$$\sum_{n=1}^{\infty} d_n^2 < \infty. \tag{8.29}$$

The treatment of the Neuman problem is similar. Let the space of admissible functions be given again by equations (8.11a). Then equation (8.9b) is replaced by

$$\frac{\partial u}{\partial n} = g_{\partial\Omega}, \quad \text{on} \quad \partial\Omega \tag{8.30}$$

where the boundary values $g_{\partial\Omega} \in \{1\}^{\perp} \subset H^{\circ}(\partial\Omega)$. The previous argument still holds if (8.17) is replaced by

$$M_{nm} = \int_{\partial\Omega} \frac{\partial w_n}{\partial n} \frac{\partial w_m}{\partial n} \, d\mathbf{x} \tag{8.31a}$$

and

$$c_m = \int_{\partial\Omega} g_{\partial\Omega} \frac{\partial w_m}{\partial n} \, d\mathbf{x}. \tag{8.31b}$$

In this case $u^N \to u$ in $H^{3/2}(\Omega)$; therefore, also $u^N \to u$ in $H^{1/2}(\Omega)$. It must be observed that this assertion is not strictly true because the solution of Neuman's problem contains an undetermined constant. To remove it one can take $\mathfrak{B} = \{1, w_1, w_2, \ldots\} \subset N^{3/2}(\Omega)$ and require

$$\int_{\partial\Omega} w_j \, d\mathbf{x} = 0, \quad j = 1, 2, \ldots. \tag{8.32}$$

Then

$$\int_{\partial\Omega} u \, d\mathbf{x} = 0. \tag{8.33}$$

In general, if the normal derivative $(\partial u^N/\partial n) \to g_{\partial\Omega}$ in $H^\circ(\partial\Omega)$, then $u^N \to u$ in $H^{3/2}(\Omega)$; hence, on the boundary $u^N \to u$ in $H^1(\partial\Omega)$, which implies $u^N \to u$ in $H^\circ(\partial\Omega)$. Thus, the boundary values (i.e., $\gamma_0 u$ on $\partial\Omega$), which in the case of the Neuman problem are not known beforehand, can be derived directly from the approximating sequence. However, the use of the reciprocity relation (8.20) offers an alternative for computing them. Indeed, one simply has to replace equations (8.21) and (8.27) by

$$u^N = \sum_{n=1}^{N} b_n^N \frac{\partial w_n}{\partial n} \to u, \quad \text{in} \quad H^\circ(\partial\Omega), \tag{8.34}$$

$$K_{nm} = \int_{\partial\Omega} \frac{\partial w_n}{\partial n} \frac{\partial w_m}{\partial n} \, \mathbf{dx} \tag{8.35a}$$

and

$$d_m = \int_{\partial\Omega} u \frac{\partial w_m}{\partial n} \, \mathbf{dx} = \int_{\partial\Omega} g_{\partial\Omega} w_m \, \mathbf{dx}. \tag{8.35b}$$

Again, equations (8.26) have to be satisfied. When this is the case, the solution u in (8.34) fulfils (8.33). This method can be used to accelerate the convergence of the approximating sequence on the boundary. As a matter of fact, when the system of equations (8.12), (8.16) and (8.31) is applied, the norm $\|(\partial u^N/\partial n) - g_{\partial R}\|$, in the $\mathcal{L}^2(\partial\Omega)$ sense, is minimal; however, $\|u^N - u\|$ in $\mathcal{L}^2(\partial\Omega)$ in general is not minimal. When equations (8.26), (8.34) and (8.35) are applied, on the contrary, $\|u^N - u\|$, in the $\mathcal{L}^2(\partial\Omega)$ sense, is minimal; i.e., in the first case the approximation of the boundary data is optimal, whereas by the second method the approximation of the unknown boundary values is optimal. In applications, generally, the latter is preferred.

Generally, when dealing with partial differential equations only some boundary values of the functions and their derivatives are relevant in the discussion of the problems. For example, for the Laplace equation these are the function u and its normal derivative $\partial u/\partial n$; for elasticity, the displacements $\mathbf{u}$ and tractions $\mathbf{T}(\mathbf{u})$. When a boundary value problem is formulated, only one part of this boundary information is prescribed and the other part must be derived after the solution has been obtained. For the Dirichlet problem, for example, u is prescribed and $\partial u/\partial n$ is derived. The converse has to be done in the case of Neuman problem. Approximating sequences for the complementary boundary values which depend on reciprocity relations, such as (8.20), can be derived for very general classes of differential equations. The reciprocity relations can be obtained from corresponding Green's formulas. For example, from

$$\int_{\Omega} \{v\Delta u - u\Delta v\}\, d\mathbf{x} = \int_{\partial\Omega} \left\{v\frac{\partial u}{\partial n} - u\frac{\partial v}{\partial n}\right\} d\mathbf{x} \tag{8.36}$$

one gets

$$\int_{\partial\Omega} v\frac{\partial u}{\partial n}\, d\mathbf{x} = \int_{\partial\Omega} u\frac{\partial v}{\partial n}\, d\mathbf{x}, \tag{8.37}$$

when u and v are harmonic in Ω. Equation (8.37) can be recognized as (8.20). The abstract formulation of Green's formulas carried out in Part 1 can be used to obtain very general reciprocity relations of this kind.

The procedure used to derive approximations (8.21) and (8.34), can be traced to a group of Italian mathematicians [73, 74, 76, 83] and is discussed extensively by Kupradze [75]. The abstract formulation which has been introduced by the author permits extension of this procedure to many problems, including those with prescribed jumps (applications to elasticity are given in [40]). This is linked to the systematic classification of boundary values associated with canonical decompositions. Generally, when a boundary problem is formulated and a Green's formula is available, Bu is prescribed and the complementary boundary values B^*u must be obtained as part of the solution.

As mentioned in the Introduction, there are several theoretical questions which have to be analyzed in order to increase the flexibility and versatility of the procedure. We will refer only to:

(a) algorithms for computing the solution in the region and on its boundary;
(b) conditions under which they converge;
(c) criteria for the completeness of a system of solutions; and
(d) change of boundary conditions.

Regarding (a), two algorithms have been presented, one for computing the solution in the interior of a region and the other one on its boundary. These algorithms converge when the system $\mathfrak{B} = \{1, w_1, w_2, \ldots\} \subset N^{3/2}(\Omega)$ spans $N^{3/2}(\Omega)$. However, in applications it is frequently difficult to verify this condition in a direct manner and it is necessary to use alternative criteria; these can be established by analyzing the spaces spanned by the boundary values. For example, for the Laplace equation, given a system of functions $\mathfrak{B}=\{w_1, w_2, \ldots\}$ defined in Ω, let us denote by $\hat{w}_\alpha = [w_{\alpha 1}, w_{\alpha 2}]$ the system of traces $w_{\alpha 1}=\gamma_0 w_\alpha$ and $w_{\alpha 2}=\gamma_1 w_\alpha$. In addition

$$\hat{\mathfrak{B}}=\{\hat{w}_1, \hat{w}_2, \ldots\}, \mathfrak{B}_1=\{w_{11}, w_{21}, \ldots\} \text{ and } \mathfrak{B}_2=\{w_{12}, w_{22}, \ldots\}.$$

For example, when the region Ω is a circle (the unit circle for definiteness), by separation of variables one obtains (in polar coordinates)

$$\mathfrak{B}=\{1; r^n \cos n\theta, r^n \sin n\theta; \quad n=1, 2, \ldots\}. \tag{8.38}$$

This system is made of harmonic polynomials

$$\mathfrak{B}=\{1, x^2-y^2, xy, \ldots\}, \tag{8.39}$$

which can be recognized as Re z^n and Im z^n $(n=0, 1, \ldots)$. Setting $r=1$ in (8.38), i.e. for a unit circle, one obtains the system of traces

$$\mathfrak{B}_1=\{\cos n\theta, \sin n\theta; \quad n=0, 1, \ldots\} \tag{8.40a}$$

and

$$\mathfrak{B}_2=\{-n \sin n\theta, n \cos n\theta; \quad n=0, 1, \ldots\}. \tag{8.40b}$$

Denote by N_1 and N_2 the spaces spanned in the $\mathscr{L}^2(\partial\Omega)$ metric by the traces $\gamma_0 u$ and $\gamma_1 u$, respectively, when u ranges over $N^{3/2}(\Omega)$. Clearly, $N_1=\mathscr{L}^2(\partial\Omega)=H^\circ(\partial\Omega)$ while $N_2=\{1\}^\perp \subset \mathscr{L}^2(\partial\Omega)$. Here, the orthogonal complement $\{1\}^\perp$ is taken in the $\mathscr{L}^2(\partial\Omega)$ inner product. Let $\mathfrak{B}\subset N^{3/2}(\Omega)$ be a system such that

$$\operatorname{span} \mathfrak{B}_1=N_1=\mathscr{L}^2(\partial\Omega) \quad \text{and} \quad \operatorname{span} \mathfrak{B}_2=N_2=\{1\}^\perp, \tag{8.41}$$

where the spans are taken in the $\mathscr{L}^2(\partial\Omega)$ sense.

For simplicity, assume that the constant function $w_0=1$ is a member of $\mathfrak{B}$, so that

$$\mathfrak{B}=\{1\}\cup\mathfrak{B}', \tag{8.42}$$

where $\mathfrak{B}'=\{w_1, w_2, \ldots\}$. It will also be assumed that

$$\int_{\partial\Omega} w_\alpha \, d\mathbf{x}=0, \quad \alpha=1, 2, \ldots. \tag{8.43}$$

Any harmonic function $u\in N^{3/2}(\Omega)$ can be written uniquely as

$$u=a_0+u', \tag{8.44}$$

where $u' \in N^{3/2}$, and α_0 is the constant:

$$\alpha_0 = \int_{\partial\Omega} u \, d\mathbf{x} \quad \text{and} \quad \int_{\partial\Omega} u' \, d\mathbf{x} = 0. \tag{8.45}$$

In view of (8.41) and $\gamma_1 w_0 = 0$, it is clear that

$$\operatorname{span} \mathfrak{B}_2' = \{1\}^{\perp}. \tag{8.46}$$

Also, $\gamma_1 u' \in \{1\}^{\perp}$, since u' is harmonic in Ω, so that $\gamma_1 u'$ is in the $\mathscr{L}^2(\Omega)$-span of $\mathfrak{B}_2'$. This shows that there is a sequence v^N of linear combinations of $\mathfrak{B}'$ such that

$$\gamma_0 v^N \xrightarrow[N\to\infty]{} \gamma_0 u', \quad \text{in} \quad \mathscr{L}^2(\partial\Omega). \tag{8.47}$$

In view of (8.43), the second of conditions (8.45) and continuity properties [82] of solutions of elliptic equations, it is clear that $v^N \to u'$ in the metric of $H^{3/2}(\Omega)$. Therefore, the linear combination $u^N = \alpha_0 + v^N$ of elements of $\mathfrak{B} \subset N^{3/2}(\Omega)$ is such that $u^N \to u$ in $H^{3/2}(\Omega)$. This shows that

$$\operatorname{span} \mathfrak{B} = N^{3/2}(\Omega), \tag{8.48}$$

where the span is taken in the $H^{3/2}(\Omega)$ metric. Thus, in this case we have derived the completeness of the system $\mathfrak{B} \subset N^{3/2}(\Omega)$ from the fact that the system of traces $\mathfrak{B}_1$ spans the same space as the traces of harmonic functions (i.e., solutions of the homogeneous equation) in $N^{3/2}(\Omega)$. Similar results hold in a more general context.

Let $\hat{\mathfrak{H}} = H^{\circ}(\partial\Omega) \oplus H^{\circ}(\partial\Omega)$ be the space of pairs $\hat{u} = [u_1, u_2]$ with $u_1 \in H^{\circ}(\partial\Omega)$ and $u_2 \in H^{\circ}(\partial\Omega)$, provided with the usual inner product

$$((\hat{u}, \hat{v})) = (u_1, v_1)_{\partial} + (u_2, v_2)_{\partial}. \tag{8.49}$$

Denote by $\hat{\mathfrak{N}} \subset \hat{\mathfrak{H}}$ the image of $N^{3/2}(\Omega)$ under the mapping $u \to \hat{u} = [\gamma_0 u, \gamma_1 u] \in \hat{\mathfrak{H}}$. It can be seen that $\hat{\mathfrak{N}} \subset \hat{\mathfrak{H}}$ is closed in the metric of $\hat{\mathfrak{H}}$. Notice that the reciprocity relation (8.20) becomes (in this chapter we assume that Hilbert spaces are being taken with real coefficients, but later this limitation will be removed)

$$(v_1, u_2) = (v_2, u_1) \quad \forall\, \hat{u} \in \hat{\mathfrak{N}} \quad \& \quad \hat{v} \in \hat{\mathfrak{N}}. \tag{8.50}$$

A system of functions $\mathfrak{B} \subset N^{3/2}(\Omega)$ will be said to be T-complete if, for every $\hat{u} \in \hat{\mathfrak{H}}$ one has

$$(w_1, u_2) = (w_2, u_1) \quad \forall\, \hat{w} \in \hat{\mathfrak{B}} \Rightarrow \hat{u} \in \hat{\mathfrak{N}}. \tag{8.51}$$

Using this notation the following characterization of complete systems holds (see Chapter 11):

Theorem 8.1. *Let* $\mathfrak{B} \subset N^{3/2}(\Omega)$ *and* $1 \in \mathfrak{B}$. *Then the following assertions are equivalent:*

(i) $\mathfrak{B} \subset N^{3/2}(\Omega)$ *spans* $N^{3/2}(\Omega)$ *in the metric of* $H^{3/2}(\Omega)$;
(ii) $\mathfrak{B} \subset \mathfrak{H}$ *spans* $\mathfrak{N}$ *in the metric of* $\mathfrak{H}$,
(iii) $\mathfrak{B} \subset \mathfrak{N}$ *is a T-complete system*;

In addition, when any of these conditions holds, equations (8.41) *are satisfied when the spans are taken in the* $\mathscr{L}^2(\partial\Omega)$ *sense.*

An advantage of having a system which satisfies any of the criteria (i) to (iii) is that the same system can be used for both a Dirichlet and a Neuman problem. Indeed, the same T-complete system can be used for general linear boundary conditions which are prescribed pointwise.

It is interesting to observe that it is possible to develop systems which are complete in regions which are, to a large extent, arbitrary. For example, the system of harmonic polynomials given by (8.38) and (8.39) is T-complete in any bounded and simply connected region (see, for example, [25]). Also, the system

$$\{\text{Log } r, \text{Re } z^{-n}, \text{Im } z^{-n}, \quad n = 1, 2, \ldots\} \tag{8.52}$$

is T-complete in the exterior of any simply connected and bounded region which contains the origin. The same is true of most systems discussed in Chapter 11.

It is of value to develop general criteria establishing conditions under which a system which is complete in one region is also complete in another, especially if such criteria are applicable to a wide class of partial differential equations. The notion of T-completeness is useful in this respect.

9 The subspace I_P

For the theory to be developed, it is not essential that the linear subspace N_P be associated with an operator $P: D \to D^*$. However, there is a property that is enjoyed by the null subspace of any operator $P: D \to D^*$, that will be required in what follows.

Lemma 9.1. *Given $P: D \to D^*$, define $A = P - P^*$. Let N_P be the null subspace of P. Then, N_P is a commutative subspace of $A: D \to D^*$.*

Proof. This follows because

$$\langle Au, v\rangle = \langle Pu, v\rangle - \langle Pv, u\rangle = 0 \tag{9.1}$$

when $u \in N_P$ and $v \in N_P$.

This is the property that will be needed. Thus, in what follows it will be assumed that a skew-symmetric operator $A: D \to D^*$ is given and $N_P \subset D$ will be assumed to be an isotropic or commutative subspace for A. The existence of an operator $P: D \to D^*$, such that N_P is the null subspace of P, will not be required. In addition, throughout this chapter $I \subset D$ will be assumed to be a regular subspace for A.

Definition 9.1. *Given $U \in D$ and $V \in D$, an element $u \in D$ is said to be a solution of the abstract boundary value problem when*

$$u - U \in N_P \quad \textit{and} \quad u - V \in I. \tag{9.2}$$

Definition 9.2. *Given the isotropic subspace N_P, define*

$$I_P = N_A + N_P. \tag{9.3}$$

Using Definition 4.1, one can give the following interpretation to Definition 9.2.

Lemma 9.2. *An element $u \in D$ belongs to I_P if and only if there exists $u_P \in N_P$ such that u and u_P assume the same relevant boundary values.*

Proof. Let $u \in I_P$, then $u = u_P + u_A$, with $u_P \in N_P$ and $u_A \in N_A$. Hence

$$Au = Au_P, \tag{9.4}$$

i.e., u and u_P have the same relevant boundary values, by virtue of Definition 4.1. Conversely, when (9.4) holds, then $u - u_P \in N_A$ and $u = u_P + (u - u_p) \in I_P$.

Example 9.1. Take D and N_P as in Chapter 8 (equations 8.11). Let

$$\langle Au, v \rangle = \int_{\partial\Omega} \left\{ u \frac{\partial v}{\partial n} - v \frac{\partial u}{\partial n} \right\} d\mathbf{x}. \tag{9.5}$$

Then

$$N_A = \left\{ u \in D \mid u = \frac{\partial u}{\partial n} = 0, \quad \text{on} \quad \partial\Omega \right\} \tag{9.6}$$

and the subspace I_P is made by the functions of D, which take the same relevant boundary values as harmonic functions of D. More specifically, $u \in D$ belongs to I_P if and only if there exists a harmonic function $u_P \in D$ such that

$$u = u_P \quad \text{and} \quad \frac{\partial u}{\partial n} = \frac{\partial u_P}{\partial n}, \quad \text{on} \quad \partial\Omega. \tag{9.7}$$

Example 9.2. For elasticity, it is similar. We do not specify precisely the space D. It can be taken as in Section 6.4 or a modification similar to that implied by equation (8.11a) can be introduced. For static problems, one defines $N_P \subset D$ as the linear space

$$N_P = \{\mathbf{u} \in D \mid \mathcal{L}_i(\mathbf{u}) = 0\}, \tag{9.8}$$

where $\mathcal{L}_i(u)$, given by (6.50) is interpreted in the sense of distributions, if necessary. Define $A : D \to D^*$ by (6.52). Then

$$N_A = \{\mathbf{u} \in D \mid \mathbf{u} = \mathbf{T}(\mathbf{u}) = 0, \quad \text{on} \quad \partial\Omega\} \tag{9.9}$$

Then $\mathbf{u} \in I_P$ if and only if there is a solution $\mathbf{u}_P \in N_P$ of the equations of elastostatics (i.e., an equilibrium state) whose displacements $\mathbf{u}_P$ and tractions $\mathbf{T}(\mathbf{u}_P)$ on the boundary are the same as those associated with the displacement field $\mathbf{u}$.

Lemma 9.3. *When N_P is isotropic, the linear space I_P defined by equation* (9.3) *is a regular subspace.*

Proof. By definition, condition (4.4) is clear. In order to show that I_P is isotropic, given $u \in I_P$ and $v \in I_P$, let $u_P \in N_P$ and $v_P \in N_P$ assume the same relevant boundary values as u and v, respectively. Then

$$\langle Au, v\rangle = \langle Au_P, v\rangle = -\langle Av, u_P\rangle = \langle Av_P, u_P\rangle = 0. \quad (9.10)$$

Definition 9.3. *The abstract boundary value problem satisfies*

(a) *existence, when there is at least one solution for every $U \in D$ and $V \in D$;*
(b) *uniqueness, when $U = 0$ and $V = 0 \Rightarrow u = 0$;*
(c) *uniqueness on the boundary, when $U = 0$ & $V = 0 \Rightarrow u \in N_A$.*

By a boundary solution is meant an element $u \in D$ such that $u - U \in I_P$ while $u - V \in I$.

Clearly any strict solution is a boundary solution. For the case when $I \subset D$ is regular, as we have assumed, a weak version of the converse statement is also true. The following lemma gives a precise formulation of this assertion.

Lemma 9.4. *An element $u \in D$ is a boundary solution of the abstract boundary value problem if and only if there exists a strict solution $u' \in D$, such that u and u' take the same relevant boundary values (i.e., $u - u' \in N_A$, according to Definition* 4.1).

Proof. This can be seen observing that, if $u \in D$ is a boundary solution, $u - U \in I_P$, which implies

$$u - U = u_P + u_A, \quad u_P \in N_P, \quad u_A \in N_A, \quad (9.11)$$

by virtue of equation (9.3). Define $u' = u - u_A$. Then it is easy to verify that u' is a solution of the abstract boundary value problem and $u - u' = u_A \in N_A$.

Lemma 9.5. *The abstract boundary value problem satisfies exis-*

tence, uniqueness or uniqueness on the boundary if and only if the problem

$$u-U\in N_P \quad \text{and} \quad u\in I, \tag{9.12}$$

or alternatively

$$u\in N_P \quad \text{and} \quad u-V\in I, \tag{9.13}$$

enjoys the corresponding property.

Proof. Let us prove the assertion of the lemma with respect to (9.12). This follows from the fact that if $w\in D$ is defined by $w=u-V$, with $V\in D$ fixed, then

$$u-U\in N_P \quad \text{and} \quad u-V\in I \Leftrightarrow w-U+V = u-U\in N_P \quad \text{and} \quad w\in I. \tag{9.14}$$

A similar argument for $u-U$ yields the other part.

Theorem 9.1. *If the abstract boundary value problem satisfies existence, then the pair $\{I, I_P\}$ constitutes a canonical decomposition of D with respect to A.*

Proof. In view of Definition 4.5 and Lemma 9.3, it is only necessary to prove that

$$D=I+I_P, \tag{9.15}$$

because both I and I_P are regular subspaces. This is immediate, because, given $u\in D$, define $u_1\in D$ by

$$u_1-u\in N_P \quad \text{while} \quad u_1\in I \tag{9.16}$$

and write $u_2=u-u_1$: then $u=u_1+u_2$, with $u_1\in I$ and $u_2\in I_P$.

Corollary 9.1. *When the abstract boundary value problem satisfies existence, I and I_P are completely regular subspaces.*

Proof. Proved by Theorems 9.1 and 4.1.

Corollary 9.2. *When the abstract boundary value problem satisfies existence, it also satisfies uniqueness on the boundary.*

Proof. This follows because

$$I\cap I_P=N_A. \tag{9.17}$$

Example 9.3. For any fixed real number s, let

$$D = \{u \in H^s(\Omega) \mid u \in H^\circ(\partial\Omega) \quad \text{and} \quad \partial u/\partial n \in H^\circ(\partial\Omega)\}, \tag{9.18}$$

$$I = \{u \in D \mid \partial u/\partial n = 0, \quad \text{on} \quad \partial\Omega\} \tag{9.19a}$$

and

$$N_P = \{u \in D \mid \Delta u + u = 0, \quad \text{in} \quad \Omega\}. \tag{9.19b}$$

Here, the reduced wave equation is interpreted in the sense of distributions. The subspace N_P is commutative for $A: D \to D^*$ as given by (3.16). Assume the region Ω is such that -1 is not an eigenvalue of the Laplacian for the boundary condition $\partial u/\partial n = 0$ on $\partial\Omega$. The abstract boundary value problem is the Neuman problem for the reduced wave equation. If $s \geq \frac{1}{2}$, the prescribed boundary values are the traces $\partial V/\partial n$ of functions $V \in D$, whose range is $H^\circ(\partial\Omega)$. This problem satisfies existence by well-known results about elliptic equations [82]. Therefore, I_P is completely regular by virtue of Corollary 9.1.

The characteristic feature of elements belonging to I_P is that they assume boundary values corresponding to solutions $u \in N_P$ of the homogeneous equation. Taking this into account, the property

$$\langle Au, v\rangle = 0 \quad \forall\, v \in I_P \Rightarrow u \in I_P, \tag{9.20}$$

which is satisfied when I_P is completely regular, can be interpreted as a purely algebraic characterization of boundary values of solutions of the homogeneous equation. Such characterization was suggested in an heuristic manner by Trefftz [58, 67] in connection with some specific problems. There is no need, of course, to use the whole set I_P in order to achieve this characterization. It is more convenient to use a T-complete subsystem $\mathfrak{B} \subset I_P$ which is known to exist by Lemma 4.3. The usefulness of this characterization is enhanced when $\mathfrak{B}$ is denumerable. The purely algebraic character of these concepts supplies greater flexibility to the theory and it will be useful when developing T-complete systems for arbitrary regions.

Example 9.4. For any fixed real number s, let the linear space D be given by (9.18). Define

$$N_P = \{u \in D \mid \Delta u = 0, \quad \text{in} \quad \Omega\} \tag{9.21}$$

and I_P by (9.3). Here, $A: D \to D^*$ is given by (3.16). When Ω is a circle of radius r with centre in the origin, it is easy to see that the system of functions (expressed in polar coordinates)

$$\mathfrak{B}=\{1, r^n \cos n\theta, r^n \sin n\theta; \quad n=1,2,\ldots\}\subset N_P \tag{9.22}$$

is T-complete for I_P. Indeed, assume $u \in D$ is such that

$$\langle Au, w\rangle = \int_0^{2\pi} \left\{ u\frac{\partial w}{\partial r} - w\frac{\partial u}{\partial r}\right\} r\, \mathrm{d}\theta = 0 \quad \forall\, w \in \mathfrak{B}. \tag{9.23}$$

Condition (9.23) is equivalent to

$$\int_0^{2\pi} \frac{\partial u}{\partial r}\, \mathrm{d}\theta = 0 \tag{9.24a}$$

together with

$$\begin{aligned} \int_0^{2\pi} u \cos n\theta \,\mathrm{d}\theta &= \frac{r}{n}\int_0^{2\pi} \frac{\partial u}{\partial r} \cos n\theta\, \mathrm{d}\theta; \\ \int_0^{2\pi} u \sin n\theta \,\mathrm{d}\theta &= \frac{r}{n}\int_0^{2\pi} \frac{\partial u}{\partial r} \sin n\theta\, \mathrm{d}\theta. \end{aligned} \tag{9.24b}$$

Let $u' \in N_P \subset D$, be a solution of the Neuman problem

$$\Delta u' = 0, \quad \frac{\partial u'}{\partial r} = \frac{\partial u}{\partial r} \tag{9.25}$$

such that

$$\int_0^{2\pi} u'\, \mathrm{d}\theta = \int_0^{2\pi} u\, \mathrm{d}\theta. \tag{9.26}$$

In view of restriction (9.24a), this problem possesses a solution. As a matter of fact, the boundary values $[u', \partial u'/\partial r]$ are such that $u' \in H^1(\partial\Omega) \subset H^\circ(\partial\Omega)$, while $\partial u'/\partial r \in H^\circ(\partial\Omega)$ [82]. Clearly, $u' \in N_P \subset I_P$; i.e., u' satisfies (9.23); therefore, also (9.24). These equations together with (9.26) imply that $u \equiv u'$ on $\partial\Omega$, because the system $\{1, r^n \cos n\theta, \ r^n \sin n\theta \ \ (n=1,2,\ldots)\}$ spans $H^\circ(\partial\Omega) = \mathscr{L}^2(\partial\Omega)$. Therefore

$$\langle Au, w\rangle = 0 \quad \forall\, w \in \mathfrak{B} \Rightarrow u \in I_P. \tag{9.27}$$

Notice that $\mathfrak{B} = \{1, \operatorname{Re} z^n, \operatorname{Im} z^n, \ (n=1,2,\ldots)\}$ is a system of harmonic polynomials. It can be shown (see Chapter 11) that this system is T-complete in any bounded and simply connected region of the plane. The possibility of making changes in the region of definition of such general character without losing the property of T-completeness supplies greater flexibility to this notion.

Frequently, it is preferable to restrict attention to spaces of boundary values. The notation to be used is $\mathfrak{D} = D/N_A$, $\mathfrak{I} = I/N_A$ and $\mathfrak{I}_P = I_P/N_A$. Also $\mathfrak{I}_1 = I_1/N_A$ and $\mathfrak{I}_2 = I_2/N_A$ when a canonical decomposition $\{I_1, I_2\}$ is available. Given $u \in D$ there is a unique element of $\mathfrak{D}$ associated with u; frequently, they will be represented by the same symbol unless such ambiguity leads to confusion. The same usage will be followed in connection with boundary operators. In particular, when $\mathfrak{B} \subset I_P$ is T-complete we write

$$\langle Au, w \rangle = 0 \quad \forall\, w \in \mathfrak{B} \Rightarrow u \in \mathfrak{I}_P. \tag{9.28}$$

Note, finally, that

$$\mathfrak{N}_A = N_A/N_A = \{0\}. \tag{9.29}$$

Thus, in general, the usage of gothic letters will imply that (9.29) holds.

Example 9.5. Corollary 9.1, can be used to show that the subspaces I_P associated with partial differential equations are generally completely regular. Later, in Chapter 11, stronger results will be shown. However, Corollary 9.1 already covers cases of considerable generality. The most general elliptic operator of order $2m$ that is formally symmetric is given by equations (3.18) and (3.19). Let $D = H^s(\Omega)$ with $s > 2m - \frac{1}{2}$. With reference to Section 6.1, let $P: D \to D^*$ be given by (3.19), $A = P - P^*$ and N_P be the null subspace of P. Assume that the only solution of the boundary value problem ($u \in D$)

$$\mathscr{L}u = 0 \text{ in } \Omega; \quad S_j u = 0, \quad j = 0, 1, \ldots, m-1, \quad \text{on } \partial\Omega \tag{9.30}$$

is $u \equiv 0$ in Ω. Under these assumptions the abstract boundary value problem of Definition 9.1 satisfies existence [82, pp. 188–189] if $I = N_B$. Therefore I_P is completely regular; i.e., for any elliptic equation satisfying (9.30), T-complete subsystems exist. Later, in Chapter 11, it will be shown that restriction (9.30) can be removed and the assertion is true for any elliptic equation which is formally symmetric.

10 Immersion in a Hilbert space

According to Definition 4.3, a linear subspace $I \subset D$ is completely regular when for every $u \in D$ one has

$$\langle Au, v \rangle = 0 \quad \forall\, v \in I \Leftrightarrow u \in I. \tag{10.1}$$

Frequently, in applications it is possible to embed D in a Hilbert space H. Generally, $D \subset H$ but $D \neq H$. Moreover, H can be taken so that D is dense in H. Concerning such embedding there are several questions which are relevant. When the embedding is carried out, the property of a given subspace of being completely regular is not necessarily preserved; more precisely, there are linear subspaces $I \subset D \subset H$ which do not satisfy (10.1) for every $u \in H$, although they do satisfy that relation for every $u \in D$. Thus, it would be interesting to find conditions under which this is preserved.

Also, it will be seen that such immersion in a Hilbert space is usually made with reference to a given Green's formula. As a matter of fact, the usefulness of a given embedding depends on the Green's formula to be applied and it is relevant to know what properties are preserved when one goes from one Green's formula to another. This is specially valuable when solving boundary value problems, since the specification of boundary conditions is usually associated with such formulas. The first of these questions has received some attention in previous work, but further research must be made to obtain definite results.

In this chapter we assume that $\mathfrak{E}$ is a separable Hilbert space with inner product $(\ ,\ )$. In addition, the antisymmetric operator $A : \mathfrak{E} \to \mathfrak{E}^*$ will be such that its null subspace

$$N_A = \{0\}. \tag{10.2}$$

If, for every $u \in \mathfrak{E}$, $Au \in \mathfrak{E}^*$ is continuous, there exists an antilinear mapping $\mathscr{A} : \mathfrak{E} \to \mathfrak{E}$ (not necessarily continuous) such that

$$\langle Au, v \rangle = (v, \mathscr{A}u). \tag{10.3}$$

When the coefficients of the linear space are complex valued, $\mathscr{A}$ is necessarily antilinear (i.e., conjugate linear [102]) and the order of the arguments in the right-hand side of equation (10.3) is relevant. Otherwise the mapping is linear and that order is immaterial.

Example 10.1. Restrict attention to the case of real coefficients. If $D = H^s(\Omega)$, $s \geq 2$, and $A : D \to D^*$ is given by

$$\langle Au, v\rangle = \int_{\partial\Omega} \left\{ v\frac{\partial u}{\partial n} - u\frac{\partial v}{\partial n} \right\} \mathrm{d}\mathbf{x}, \tag{10.4}$$

let $\mathfrak{E} = \mathfrak{D} = D/N_A$, where $\mathfrak{E} = H^{s-(1/2)}(\partial\Omega) \oplus H^{s-(3/2)}(\partial\Omega)$. In this case every element $\hat{u} \in \mathfrak{E}$ is characterized by the pair of traces $[u, \partial u/\partial n]$. Then, for every fixed $u \in \mathfrak{E}$, $Au \in \mathfrak{E}^*$ is continuous and there exists $\mathscr{A}u \in \mathfrak{E}$ which satisfies (10.3). Thus

$$(\mathscr{A}u, v) = (p_1, v)_{s-(1/2)} - (p_2, \partial v/\partial n)_{s-(3/2)}, \tag{10.5}$$

where $p_1 \in H^{s-(1/2)}(\partial\Omega)$ and $p_2 \in H^{s-(3/2)}(\partial\Omega)$ are such that

$$(p_1, v)_{s-(1/2)} = \int_{\partial\Omega} v\frac{\partial u}{\partial n}\,\mathrm{d}\mathbf{x}, \quad (p_2, \partial v/\partial n)_{s-(3/2)} = \int_{\partial\Omega} u\frac{\partial v}{\partial n}\,\mathrm{d}\mathbf{x}. \tag{10.6}$$

Here $(\,,\,)_{s-(1/2)}$ and $(\,,\,)_{s-(3/2)}$ stand for the inner products in $H^{s-(1/2)}(\partial\Omega)$ and $H^{s-(3/2)}(\partial\Omega)$, respectively. This result is unnecessarily complicated and not suitable for most applications. The structure, however, will be simplified by results to be presented in this and the next chapter.

Lemma 10.1. *Let $\mathfrak{I}_0 \subset \mathfrak{E}$ be a linear subspace (not necessarily closed). Then, the following statements are equivalent*

(i) *$\mathfrak{I}_0$ is completely regular;*

(ii) $\mathfrak{I}_0 = (\mathscr{A}\mathfrak{I}_0)^\perp$. (10.7)

Here the orthogonal complement is taken in $\mathfrak{E}$.

Proof. This follows because (i) means

$$(v, \mathscr{A}u)_0 = \langle Au, v\rangle = 0 \quad \forall\, v \in \mathfrak{I}_0 \Leftrightarrow u \in \mathfrak{I}_0.$$

This is clearly equivalent to (10.7). Notice that Lemma 10.1 is a generalization of property (iii) obtained in Example 4.6.

Corollary 10.1. *A completely regular subspace $\mathfrak{I}_0 \subset \mathfrak{E}$ is necessarily closed.*

Proof. This follows because the orthogonal complement of any subspace is closed.

The relation

$$(u, \mathscr{A}v) = -(v, \mathscr{A}u) \tag{10.8}$$

will be used in the sequel. It follows from

$$(v, \mathscr{A}u) = \langle Au, v\rangle = -\langle Av, u\rangle = -(u, \mathscr{A}v). \tag{10.9}$$

Lemma 10.2. *For every $u \in \mathfrak{E}$ and $v \in \mathfrak{E}$, the relations*

(i) $(\mathscr{A}^2 u, v) = -(\mathscr{A}v, \mathscr{A}u) = (u, \mathscr{A}^2 v)$ (10.10)

and

(ii) $-(\mathscr{A}^2 u, u) = (\mathscr{A}u, \mathscr{A}u) = \|\mathscr{A}u\|^2$ (10.11)

holds. Thus the operator $-\mathscr{A}^2 : \mathfrak{E} \to \mathfrak{E}$ is positive definite and self-adjoint.

Proof. Observe that

$$(\mathscr{A}^2 u, v) = \overline{(v, \mathscr{A}^2 u)} = -\overline{(\mathscr{A}u, \mathscr{A}v)} = -(\mathscr{A}v, \mathscr{A}u). \tag{10.12}$$

Here the bar stands for the complex conjugate. Replacing v by u in (10.12), one obtains (10.11). The positive definiteness of $-\mathscr{A}^2$ is clear because $\mathscr{A}u \neq 0$ whenever $u \neq 0$, by virtue of (10.2). Also

$$(u, \mathscr{A}^2 v) = \overline{(\mathscr{A}^2 v, u)} = -(\mathscr{A}v, \mathscr{A}u).$$

This equation yields (10.10) by virtue of (10.12).

Theorem 10.1. *Assume that* (10.2) *holds and $\mathscr{A} : \mathfrak{E} \to \mathfrak{E}^*$ is continuous. Then, there exists a subspace $I_0 \subset D = \mathfrak{E}$, which is completely regular for A.*

Proof. By spectral theorems there is an orthogonal basis $\{\psi_1, \psi_2, \ldots\}$ such that

$$\mathscr{A}^2 \psi_n = -\mu_n^2 \psi_n, \tag{10.13}$$

where $\mu_n \neq 0$ is real. Observe that for every ψ_n one has

$$\mathscr{A}^2(\mathscr{A}\psi_n) = -\mu_n^2 \mathscr{A}\psi_n. \tag{10.14}$$

Moreover,

$$(\psi_n, \mathscr{A}\psi_n) = \langle A\psi_n, \psi_n\rangle = 0. \tag{10.15}$$

Using these remarks, it is easy to see that one can choose an orthonormal system $\{\phi_1, \phi_2, \ldots\}$ so that the system

$$\{\phi_1, \phi_2, \ldots\} \cup \{\mathcal{A}\phi_1, \mathcal{A}\phi_2, \ldots\} \tag{10.16}$$

is an orthogonal (but not necessarily normal) system of eigenvectors.

This can be established using the fact that if $\phi_i \perp \phi_j$ then $\mathcal{A}\phi_i \perp \mathcal{A}\phi_j$. Indeed, if $\phi_i \perp \phi_j$, then

$$(\mathcal{A}\phi_j, \mathcal{A}\phi_i) = -(\phi_i, \mathcal{A}^2\phi_j) = \mu_j^2(\phi_i, \phi_j) = 0.$$

Let

$$I_0 = \text{span}\,\{\phi_1, \phi_2, \ldots\}. \tag{10.17}$$

Then (10.7) is fulfilled because

$$\mathcal{A}I_0 = \text{span}\,\{\mathcal{A}\phi_1, \mathcal{A}\phi_2, \ldots\} \tag{10.18}$$

is orthogonal to I_0, since $\phi_i \perp \mathcal{A}\phi_j$ for every $i = 1, 2, \ldots$, and $j = 1, 2, \ldots$. Clearly, the pair of subspaces $\{\mathfrak{J}_0, \mathcal{A}\mathfrak{J}_0\}$ is a canonical decomposition of $\mathfrak{E}$.

The case when all the eigenvalues $\mu_j^2 = 1$ has special interest. In what follows $(,)_{\mathfrak{E}}$ will denote the inner product in $\mathfrak{E}$. Also, when $\mathfrak{H}$ is a Hilbert space with inner product $(,)$, by a unitary antilinear mapping we mean an antilinear mapping $u \to \bar{u}$ of $\mathfrak{H}$ into $\mathfrak{H}$, such that

$$(\bar{u}, \bar{v}) = (v, u). \tag{10.19}$$

Definition 10.1. *The antisymmetric bilinear form* $A : \mathfrak{E} \to \mathfrak{E}^*$ *is said to be in standard form when*

(a) $\mathfrak{E} = \hat{\mathfrak{H}} = \mathfrak{H} \oplus \mathfrak{H}$, *where* $\mathfrak{H}$ *is a Hilbert space with inner product* $(,)$. *Thus the inner product on* $\mathfrak{E}$ *is given, for every* $\hat{u} = [u_1, u_2] \in \mathfrak{E} = \hat{\mathfrak{H}}$ *and* $\hat{v} = [v_1, v_2] \in \mathfrak{E} = \hat{\mathfrak{H}}$, *by*

$$(\hat{u}, \hat{v})_{\mathfrak{E}} = ((\hat{u}, \hat{v})) = (u_1, v_1) + (u_2, v_2). \tag{10.20}$$

Here, $((\;))$ *is the inner product on* $\hat{\mathfrak{H}}$.
(b) *There is defined on* $\mathfrak{H}$ *a unitary antilinear mapping* $u \to \bar{u}$ *such that* $\bar{\bar{u}} = u$.
(c) *For every* $\hat{u} \in \hat{\mathfrak{H}}$ *and* $\hat{v} \in \hat{\mathfrak{H}}$,

$$\langle \hat{A}\hat{u}, \hat{v} \rangle = (v_1, \bar{u}_2) - (u_1, \bar{v}_2). \tag{10.21}$$

If $\mathfrak{H}$ *is a Hilbert space with real coefficients, the mapping* $u \to \bar{u}$

must be interpreted as the identity mapping and the bar can be deleted everywhere.

Proposition 10.1. *Assume* $\hat{A}: \hat{\mathfrak{H}} \to \hat{\mathfrak{H}}$ *is in standard form. Define* $J: \hat{\mathfrak{H}} \to \hat{\mathfrak{H}}$ *by*†

$$J\hat{u} = [\bar{u}_2, -\bar{u}_1]. \tag{10.22}$$

Then:

(i) $N_{\hat{A}} = 0.$ (10.23)

(ii) $\langle \hat{A}\hat{u}, \hat{v} \rangle = ((\hat{v}, J\hat{u})).$ (10.24)

(iii) *The transformation* $J: \hat{\mathfrak{H}} \to \hat{\mathfrak{H}}$ *is unitary and antilinear; i.e. for every complex number* a, *one has*

$$J(a\hat{u}) = \bar{a}J(\hat{u}) \tag{10.25a}$$

and

$$((\hat{u}, \hat{v})) = ((J\hat{v}, J\hat{u})). \tag{10.25b}$$

(iv) $J^2 = -1.$ (10.26)

(v) *The pair of subspaces* $\{\hat{\mathfrak{I}}_1, \hat{\mathfrak{I}}_2\}$ *defined by*

$$\hat{\mathfrak{I}}_1 = \{[u_1, 0] \in \hat{\mathfrak{H}}\}, \quad \hat{\mathfrak{I}}_2 = \{[0, u_2] \in \hat{\mathfrak{H}}\} \tag{10.27}$$

is a canonical decomposition of $\hat{\mathfrak{H}}$ *with respect to* $\hat{A}: \hat{\mathfrak{H}} \to \hat{\mathfrak{H}}^*$.

Proof. It is straightforward. Indeed, $\langle A\hat{u}, \hat{v} \rangle = 0$ for every $\hat{v} \in \hat{\mathfrak{H}}$ implies $\bar{u}_2 = u_1 = 0$, by virtue of (10.21). Thus (10.23) is clear. Properties (ii) to (iv) can be verified using equations (10.21) and (10.22). In view of equation (10.23), to prove (v) it is enough to show that $\hat{\mathfrak{I}}_1$ and $\hat{\mathfrak{I}}_2$ are commutative, and that

$$\hat{\mathfrak{H}} = \hat{\mathfrak{I}}_1 + \hat{\mathfrak{I}}_2. \tag{10.28}$$

Both of these facts are clear.

Under very general conditions, it is possible to imbed antisymmetric bilinear forms in standard form. This yields, by the way, connections between the algebraic theory presented in this book and the one developed by Arens, Coddington and Lee [2–12]. A few specific examples will be given in the remainder of this chapter and a more general discussion will follow in Chapter 11.

† This notation is based on that used by Coddington and Lee [3–12].

Example 10.2. The procedure used in Example 10.1 to carry out the immersion in a Hilbert space is too complicated. Also, $\mathscr{A}:\mathfrak{E}\to\mathfrak{E}$ is not bicontinuous. A simpler procedure will be explained here. Let $D=H^{3/2}(\Omega)$, where the Hilbert space $H^{3/2}(\Omega)$ is taken with complex coefficients. The boundary values D/N_A, when A is given by (10.4), can be identified with $H^{\circ}(\partial\Omega)\oplus H^1(\partial\Omega)$, which in turn can be densely imbedded in $H^{\circ}(\partial\Omega)\oplus H^{\circ}(\partial\Omega)$. Indeed, let $\mathfrak{H}=H^{\circ}(\partial\Omega)$, then every element $\hat{u}\in D/N_A$ can be identified with a pair $\hat{u}=[u_1, u_2]$, $u_1\in H^{\circ}(\partial\Omega)\subset\mathfrak{H}$ and $u_2\in H^1(\partial\Omega)\subset\mathfrak{H}$, if one defines

$$u_1=-\frac{\partial u}{\partial n} \quad \text{and} \quad u_2=u, \quad \text{on} \quad \partial\Omega, \tag{10.29}$$

Then $D/N_A\subset\hat{\mathfrak{H}}=\mathfrak{H}\oplus\mathfrak{H}$ and the inner product on $\hat{\mathfrak{H}}$ is

$$((\hat{u}, \hat{v}))=\int_{\partial\Omega}\{u_1\bar{v}_1+u_2\bar{v}_2\}\,\mathrm{d}\mathbf{x}. \tag{10.30}$$

Here the unitary antilinear mapping $u\to\bar{u}$ is that which associates with every function $u\in\mathfrak{H}=H^{\circ}(\partial\Omega)$ its complex conjugate. The canonical decomposition of $\hat{\mathfrak{H}}$ associated with $\{\hat{\mathfrak{I}}_1, \hat{\mathfrak{I}}_2\}$ as given by (10.27) is related with the one corresponding to $\{I_1, I_2\}$ in Example 4.4.

Example 10.3. Similarly, when the biharmonic equation is considered

$$\langle Au, v\rangle=\int_{\partial\Omega}\left\{v\frac{\partial\Delta u}{\partial n}-\Delta u\frac{\partial v}{\partial n}+\Delta v\frac{\partial u}{\partial n}-u\frac{\partial\Delta u}{\partial n}\right\}\mathrm{d}\mathbf{x}. \tag{10.31}$$

The basic linear space may be taken as $H^{7/2}(\Omega)$. Then every element of the space D/N_A of boundary values can be characterized by four functions:

$$u^{\mathrm{i}}=u, \quad u^{\mathrm{ii}}=\partial u/\partial n, \quad u^{\mathrm{iii}}=\partial\Delta u/\partial n, \quad u^{\mathrm{iv}}=-\Delta u, \quad \text{on } \partial\Omega. \tag{10.32}$$

Thus, D/N_A can be identified with $H^3(\partial\Omega)\oplus H^2(\partial\Omega)\oplus H^{\circ}(\partial\Omega)\oplus H^1(\partial\Omega)$, which in turn can be densely imbedded in $\hat{\mathfrak{H}}=\mathfrak{H}\oplus\mathfrak{H}$, if $\mathfrak{H}=H^{\circ}(\partial\Omega)\oplus H^{\circ}(\partial\Omega)$. To be specific, the latter immersion will be made, defining

$$[u^{\mathrm{i}}, u^{\mathrm{ii}}]=u_1\in\mathfrak{H} \quad \text{and} \quad [u^{\mathrm{iii}}, u^{\mathrm{iv}}]=u_2\in\mathfrak{H}. \tag{10.33}$$

If $J:\hat{\mathfrak{H}}\to\hat{\mathfrak{H}}$ is given by (10.22), then equation (10.24) is fulfilled, because the inner product $((\,,\,))$ in $\hat{\mathfrak{H}}$, is given by

$$((\hat{u}, \hat{v})) = \int_{\partial\Omega} \{u^{\mathrm{i}}\bar{v}^{\mathrm{i}} + u^{\mathrm{ii}}\bar{v}^{\mathrm{ii}} + u^{\mathrm{iii}}\bar{v}^{\mathrm{iii}} + u^{\mathrm{iv}}\bar{v}^{\mathrm{iv}}\}\,\mathrm{d}\mathbf{x}. \tag{10.34}$$

Again, the unitary antilinear mapping $u \to \bar{u}$ is obtained taking the complex conjugate of the functions. Let $I_1 \subset D$ and $I_2 \subset D$ be

$$I_1 = \{u \in D \mid \Delta u = \partial\Delta u/\partial n = 0, \quad \text{on} \quad \partial\Omega\} \tag{10.35a}$$

and

$$I_2 = \{u \in D \mid u = \partial u/\partial n = 0, \quad \text{on} \quad \partial\Omega\} \tag{10.35b}$$

Then the canonical decomposition $\{\hat{\mathfrak{I}}_1, \hat{\mathfrak{I}}_2\}$ defined by (10.27) has the property

$$\hat{\mathfrak{I}}_1 \supset I_1/N_{\mathrm{A}}, \quad \hat{\mathfrak{I}}_2 \supset I_2/N_{\mathrm{A}}. \tag{10.36}$$

Example 10.4. For problems of potential theory and reduced wave equation, with prescribed jump conditions (Section 7.1), one can take $D_R = H^{3/2}(R)$ and $D_E = H^{3/2}(E)$. For simplicity, set $k_R = k_E = 1$. Then

$$\langle \hat{A}\hat{u}, \hat{v}\rangle = \int_{\partial_2 R} \left\{[\hat{u}]\left(\frac{\partial \dot{v}}{\partial n}\right) - (\dot{v})\left[\frac{\partial \hat{u}}{\partial n}\right] - [\hat{v}]\left(\frac{\partial \dot{u}}{\partial n}\right) + (\dot{u})\left[\frac{\partial \hat{v}}{\partial n}\right]\right\} \mathrm{d}\mathbf{x}. \tag{10.37}$$

Elements of the linear space $\hat{D}/N_{\mathrm{A}}$ (where $\hat{D} = D_R \oplus D_E$) can be characterized by four functions,

$$u^{\mathrm{i}} = [\hat{u}], \quad u^{\mathrm{ii}} = \left[\frac{\partial \hat{u}}{\partial n}\right], \quad u^{\mathrm{iii}} = -\left(\frac{\partial \dot{u}}{\partial n}\right), \quad u^{\mathrm{iv}} = (\dot{u}), \tag{10.38}$$

defined on $\partial_3 R$. All of them belong to $H^{\circ}(\partial_3 R)$. Thus, $\hat{D}/N_{\hat{A}}$ can be imbedded in the space $\hat{\mathfrak{H}} = \mathfrak{H} \oplus \mathfrak{H}$, where $\mathfrak{H} = H^{\circ}(\partial R) \oplus H^{\circ}(\partial R)$. For this purpose one can use equation (10.33). Again, equation (10.24) is satisfied when $\hat{J}: \hat{\mathfrak{H}} \to \hat{\mathfrak{H}}$ is given by (10.22).

The procedures that have been explained are independent of the type of equation considered. The following examples illustrate their application to parabolic and hyperbolic equations.

Example 10.5. For simplicity, attention will be restricted to the heat equation. In this case (Section 6.2),

$$\langle Pu, v\rangle = \int_{\Omega} v * \left(\frac{\partial u}{\partial t} - \Delta u\right) \mathrm{d}\mathbf{x} \tag{10.39}$$

and

$$\langle Au, v\rangle = \int_{\partial\Omega} \left\{ u * \frac{\partial v}{\partial n} - v * \frac{\partial u}{\partial n} \right\} \mathrm{d}\mathbf{x} + \int_{\Omega} \{v(0)u(T) - v(T)u(0)\} \, \mathrm{d}\mathbf{x}. \quad (10.40)$$

Functions of the space D/N_A are characterized by four functions:

$$u^{\mathrm{i}}(\mathbf{x}) = u(\mathbf{x}, 0), \quad u^{\mathrm{iii}}(\mathbf{x}) = u(\mathbf{x}, T), \quad \mathbf{x} \in \Omega, \quad (10.41a)$$

and

$$u^{\mathrm{ii}}(\mathbf{x}, t) = u(\mathbf{x}, t), \quad u^{\mathrm{iv}}(\mathbf{x}, t) = -\frac{\partial u}{\partial n}(\mathbf{x}, T-t), \quad \{\mathbf{x}, t\} \in \partial\Omega \times [0, T]. \quad (10.41b)$$

Define $\mathfrak{H} = H^{\circ}(\Omega) \oplus H^{\circ}(\partial\Omega \times [0, T])$. Then D/N_A can be imbedded in $\mathfrak{H} \oplus \mathfrak{H}$ by means of

$$u_1 = \{u^{\mathrm{i}}, u^{\mathrm{ii}}\}, \quad u_2 = \{u^{\mathrm{iii}}, u^{\mathrm{iv}}\}. \quad (10.42)$$

With this definition (10.24) is satisfied when (10.22) is adopted.

Example 10.6. For the wave equation (Section 6.3), one gets

$$\langle Pu, v\rangle = \int_{\Omega} v * \left(\frac{\partial^2 u}{\partial t^2} - \Delta u \right) \mathrm{d}\mathbf{x} \quad (10.43)$$

and

$$\langle Au, v\rangle = \int_{\Omega} \left\{ u(T) \frac{\partial v}{\partial t}(0) + \frac{\partial u}{\partial t}(T) v(0) - v(T) \frac{\partial u}{\partial t}(0) - u(0) \frac{\partial v}{\partial t}(T) \right\} \mathrm{d}\mathbf{x} + \int_{\partial\Omega} \left\{ u * \frac{\partial v}{\partial n} - v * \frac{\partial u}{\partial n} \right\} \mathrm{d}\mathbf{x}. \quad (10.44)$$

In this case $\mathfrak{H} = H^{\circ}(\Omega) \oplus H^{\circ}(\Omega) \oplus H^{\circ}(\partial\Omega \times [0, T])$. The immersion is standard if one defines

$$u^{\mathrm{i}}(\mathbf{x}) = u(\mathbf{x}, T), \quad u^{\mathrm{ii}}(\mathbf{x}) = \frac{\partial u}{\partial t}(\mathbf{x}, T), \quad \text{in} \quad \Omega, \quad (10.45a)$$

$$u^{\mathrm{iv}}(\mathbf{x}) = -\frac{\partial u}{\partial t}(\mathbf{x}, 0), \quad u^{\mathrm{v}}(\mathbf{x}) = -u(\mathbf{x}, 0), \quad \text{in} \quad \Omega, \quad (10.45b)$$

and

$$u^{\text{iii}}(\mathbf{x}, t) = u(\mathbf{x}, t), \quad u^{\text{vi}}(\mathbf{x}, t) = -\frac{\partial u}{\partial n}(\mathbf{x}, t), \quad \{\mathbf{x}, t\} \in \partial\Omega \times [0, T]. \tag{10.45c}$$

Then, the immersion

$$u_1 = [u^{\text{i}}, u^{\text{ii}}, u^{\text{iii}}], \quad u_2 = \{u^{\text{iv}}, u^{\text{v}}, u^{\text{vi}}\} \tag{10.46}$$

is standard.

11 Criterion of completeness

Throughout this chapter the notation introduced in the last one will be used. Thus, $\hat{\mathfrak{H}} = \mathfrak{H} \oplus \mathfrak{H}$ where $\mathfrak{H}$ is a Hilbert space. The criterion of completeness to be discussed is T-completeness. Recall Definition 4.4.

Definition 11.1. *A subset $\mathfrak{B} \subset \hat{\mathfrak{I}}_P \subset \hat{\mathfrak{H}}$ is said to be T-complete for $\hat{\mathfrak{I}}_P$ when, for every $\hat{u} \in \hat{\mathfrak{H}}$, one has*

$$\langle \hat{A}\hat{u}, \hat{w} \rangle = 0 \quad \forall\, \hat{w} \in \mathfrak{B} \Rightarrow \hat{u} \in \hat{\mathfrak{I}}_P. \tag{11.1}$$

The adequacy of this criterion depends on the following result.†

Proposition 11.1. *Let $\hat{A} : \hat{\mathfrak{H}} \to \hat{\mathfrak{H}}^*$ be in standard form and $\hat{\mathfrak{I}}_P \subset \hat{\mathfrak{H}}$ be completely regular. Then, a system $\mathfrak{B} = \{\hat{w}_1, \hat{w}_2, \ldots\} \subset \hat{\mathfrak{I}}_P$ is T-complete if and only if*

$$\operatorname{span} \mathfrak{B} = \hat{\mathfrak{I}}_P. \tag{11.2}$$

Proof. This follows because

$$((\hat{u}, J\hat{w}_\alpha)) = \langle \hat{A}\hat{w}_\alpha, \hat{u} \rangle = -\langle \hat{A}\hat{u}, \hat{w}_\alpha \rangle. \tag{11.3}$$

Thus, a necessary and sufficient condition for (11.1) to be satisfied is that $\{J\hat{w}_1, J\hat{w}_2, \ldots\}$ span $J(\hat{\mathfrak{I}}_P)$, which is the orthogonal complement of $\hat{\mathfrak{I}}_P$ in $\hat{\mathfrak{H}}$. Clearly, condition (11.2) is equivalent to

$$\operatorname{span} J(\mathfrak{B}) = J(\hat{\mathfrak{I}}_P), \tag{11.4}$$

because J is bicontinuous.

A corollary of Proposition 11.1 is the following.

Proposition 11.2. *Under assumptions of Proposition 11.1, when $\mathfrak{B} = \{\hat{w}_1, \hat{w}_2, \ldots\} \subset \hat{\mathfrak{I}}_P$ is T-complete for $\hat{\mathfrak{I}}_P$, one has*

$$\operatorname{span} \mathfrak{B}_1 = \mathfrak{I}_{P1}^c \quad \textit{and} \quad \operatorname{span} \mathfrak{B}_2 = \mathfrak{I}_{P2}^c. \tag{11.5}$$

† The proof of Theorem 10.1 of [35] is faulty and the result false as stated. In this chapter some related developments are presented, in which the error has already been amended.

Here, the superindex c *stands for the closure in* $\mathfrak{H}$. *Also,* $\mathfrak{B}_1 \subset \mathfrak{H}$ *and* $\mathfrak{B}_2 \subset \mathfrak{H}$ *stand for the ranges covered by the first and second components of elements of* $\mathfrak{B} \subset \hat{\mathfrak{H}}$.

Proof. This follows because (11.2) implies span $\mathfrak{B}_1 \supset \mathfrak{I}_{P1}$. Hence the first of equations (11.5), since span $\mathfrak{B}_1$ is closed. The second equation is obtained similarly.

The manner in which properties (11.2) and (11.5) can be used in specific applications can be better understood by looking again at the example considered at the beginning of Chapter 8. We look for solutions of the abstract boundary value problem (Definition 9.1), for the case when D is defined by equations (8.11), $N_P \subset D$ is the linear manifold of harmonic functions and (Equation 8.7)

$$I = \{u \in D \mid u = 0, \quad \text{on} \quad \partial\Omega\}. \tag{11.6}$$

A function $u \in D$ is a solution of this problem if and only if

$$\Delta u = f_R, \quad \text{on} \quad \Omega \tag{11.7a}$$

and

$$\gamma_0 u = \gamma_0 V, \quad \text{on} \quad \partial\Omega, \tag{11.7b}$$

where

$$f_R = \Delta U. \tag{11.8}$$

Here Δ must be interpreted in the sense of distributions. It has already been observed that $N_P \subset H^{3/2}(\Omega) \subset D$; moreover, N_P is closed (thus, it is a Hilbert space) with respect to the norm of $H^{3/2}(\Omega)$.

Clearly, any solution can be written as

$$u = U + v, \tag{11.9}$$

where $v \in N_P = N^{3/2}(\Omega)$ and $\Delta U = f_R$. Here $N^{3/2}(\Omega)$ is the Hilbert subspace of harmonic functions which belong to $H^{3/2}(\Omega)$. Let us define the mapping $\mu : D \to \hat{\mathfrak{H}}$, where $\hat{\mathfrak{H}} = \mathfrak{H} \oplus \mathfrak{H}$, $\mathfrak{H} = H^{\circ}(\partial\Omega)$ and

$$\mu(u) = \{-\gamma_1 u, \gamma_0 u\}, \tag{11.10}$$

where γ_1 and γ_0 are trace operators. Notice that this mapping corresponds to equation (10.29) which was considered in Chapter 10. Assume that $\mathfrak{B} = \{w_1, w_2, \ldots\} \subset N_P$ is such that

$$\text{span}\,\{w_1, w_2, \ldots\} = N_P \text{ (in the } H^{3/2}(\Omega) \text{ metric).} \tag{11.11}$$

Then any $v \in N_P$ can be approximated by linear combinations of $\mathfrak{B}$, i.e., such system $\mathfrak{B} \subset N_P$ can be used to construct the solution of the abstract boundary value problem. Let

$$\hat{\mathfrak{B}} = \{\hat{w}_1, \hat{w}_2, \ldots\} \subset \hat{\mathfrak{I}}_P \subset \hat{\mathfrak{H}}, \tag{11.12a}$$

where

$$\hat{w}_\alpha = \mu(w_\alpha), \quad \alpha = 1, 2, \ldots, \tag{11.12b}$$

while

$$\hat{\mathfrak{I}}_P = \hat{\mathfrak{N}}_P = \mu(N_P) = \mu(I_P). \tag{11.13}$$

Here $I_P = N_P + N_A$, with $A: D \to D^*$ given by (8.18). Define the skew-symmetric bilinear functional $\hat{A}: \hat{\mathfrak{H}} \to \hat{\mathfrak{H}}^*$ by

$$\langle \hat{A}\hat{u}, \hat{v}\rangle = (v_1, \bar{u}_2) - (u_1, \bar{v}_2). \tag{11.14}$$

Then

$$\langle Au, v\rangle = \langle \hat{A}\hat{u}, \hat{v}\rangle, \tag{11.15}$$

where $\hat{u} = \mu(u)$ and $\hat{v} = \mu(v)$. Thus, the mapping $\mu: D \to \hat{\mathfrak{H}}$ is a symplectic homomorphism between A, D and $\hat{A}, \hat{\mathfrak{H}}$. Let $\mu_r: N_P \to \hat{\mathfrak{I}}_P$ be the restriction of μ to N_P. Notice, however, that the counter domain of μ_r has also been restricted. It can be seen that

(a) $\hat{\mathfrak{I}}_P \subset \hat{\mathfrak{H}}$ is completely regular for $\hat{A}: \hat{\mathfrak{H}} \to \hat{\mathfrak{H}}^*$;
(b) the mapping $\mu_r: N_P \to \hat{\mathfrak{I}}_P$ is one-to-one. Moreover, it is a topological isomorphism;
(c) a system $\mathfrak{B} \subset N_P$ spans N_P, in the $H^{3/2}(\Omega)$ norm, if and only if $\hat{\mathfrak{B}} = \mu(\mathfrak{B}) \subset \hat{\mathfrak{I}}_P$ is T-complete for $\hat{\mathfrak{I}}_P$ with respect to $\hat{A}$; and
(d) a system $\mathfrak{B} \subset N_P$ is T-complete for $\mathfrak{I}_P$ with respect to $A: D \to D^*$, if and only if $\mathfrak{B} \subset N_P$ spans N_P in the $H^{3/2}(\Omega)$ norm.

Properties (a) to (d) follow from results we proceed to establish. Advantages of using the T-completeness criterion are due to the fact that it is easy to apply because it depends on boundary values only.

In view of the previous example, consider the following abstract situation. Let D be a linear space and $A: D \to D^*$ a skew-symmetric bilinear form. Assume $\hat{\mathfrak{H}} = \mathfrak{H} \oplus \mathfrak{H}$ where $\mathfrak{H}$ is a Hilbert space, and define $\hat{A}: \hat{\mathfrak{H}} \to \hat{\mathfrak{H}}^*$ by

$$\langle \hat{A}\hat{u}, \hat{v}\rangle = ((\hat{v}, J\hat{u})), \tag{11.16}$$

where

$$J\hat{u} = [\bar{u}_2, -\bar{u}_1]. \tag{11.17}$$

Here $\hat{u} = [u_1, u_2]$. Thus, $\hat{A}$ is in standard form. Let $\mu : D \to \hat{\mathfrak{H}}$ be a symplectic homomorphism between D, A and $\hat{\mathfrak{H}}$, $\hat{A}$. (Recall Remark 4.3.)

Remark 11.1. When $\mu : D \leftarrow \hat{\mathfrak{H}}$ is a symplectic homomorphism between D, A and $\hat{\mathfrak{H}}$, $\hat{A}$, then the mapping μ is a bijection of D/N_A into $\hat{\mathfrak{H}}$ (i.e., for every $u \in D$ and $v \in D$ one has $\mu(u) = \mu(v) \Leftrightarrow Au = Av$). This is because when $\mu : D \to \hat{\mathfrak{H}}$ is onto,

$$\langle Au, v\rangle = 0 \quad \forall\, v \in D \Leftrightarrow (v_1, \bar{u}_2) = (v_2, \bar{u}_1) = 0 \qquad \forall\, v_1 \in \mathfrak{H} \quad \& \quad v_2 \in \mathfrak{H}.$$

In what follows, $N_P \subset D$ will be assumed to be a linear subspace of D, such that N_P is a Hilbert space. However, in general there is no Hilbert space structure defined on the linear space D. Taking $I_P = N_A + N_P$, define $\hat{\mathfrak{I}}_P \subset \hat{\mathfrak{H}}$ and $\mathfrak{R}_P \subset \hat{\mathfrak{H}}$ by

$$\hat{\mathfrak{I}}_P = \mu(I_P) = \mu(N_P) = \mathfrak{R}_P. \tag{11.18}$$

This chain of equalities holds because $\mu(N_A) = \{0\}$ when $\mu : D \to \hat{\mathfrak{H}}$ is an homomorphism (thus, onto) and $\hat{A}$ is in standard form.

Lemma 11.1. *Let $\mu : D \to \hat{\mathfrak{H}}$ be a symplectic homomorphism. Then the restriction $\mu_r : N_P \to \hat{\mathfrak{I}}_P$ is one-to-one if and only if*

$$N_P \cap N_A = \{0\}. \tag{11.19}$$

Proof. For every $u \in D$ one has $\mu(u) = 0 \Leftrightarrow u \in N_A$. Using this property, the lemma is clear.

The following notation will be used. With every subset $\hat{\mathfrak{B}} \subset \hat{\mathfrak{H}}$ we associate the set of components $\mathfrak{B}_1$ and $\mathfrak{B}_2$, given by

$$\mathfrak{B}_1 = \{u_1 \in \mathfrak{H} \mid \exists\, \hat{u} = [u_1, u_2] \in \hat{\mathfrak{B}}\} \tag{11.20a}$$

and

$$\mathfrak{B}_2 = \{u_2 \in \mathfrak{H} \mid \exists\, \hat{u} = [u_1, u_2] \in \hat{\mathfrak{B}}\}. \tag{11.20b}$$

Thus, $\mathfrak{B}_1 \subset \mathfrak{H}$ and $\mathfrak{B}_2 \subset \mathfrak{H}$ can be thought of as orthogonal projections in the coordinate spaces. When $\hat{\mathfrak{I}}_P$ is a linear space, $\mathfrak{I}_{P\alpha} \subset \mathfrak{H}$ $(\alpha = 1, 2)$ are linear subspaces of $\mathfrak{H}$ and $\mathfrak{I}_{P\alpha}^{\perp}$ will be the orthogonal complement in $\mathfrak{H}$. Also, $\mathfrak{I}_{P\alpha}^{c}$ will be used for the closure of $\mathfrak{I}_{P\alpha}$ in $\mathfrak{H}$. Observe that in general $\mathfrak{I}_{P\alpha}^{c} \supset (\hat{\mathfrak{I}}_P^{c})_\alpha$, $\alpha = 1, 2$, but these sets are not necessarily equal. The former is always closed, while the latter may not be closed. Define

$$\mathfrak{N}_0^L=\{[\bar{u}_1, 0] \mid u_1\in\mathfrak{I}_{P2}^{\perp}\} \quad\text{and}\quad \mathfrak{N}_0^R=\{[0, u_2] \mid u_2\in\mathfrak{I}_{P1}^{\perp}\}. \tag{11.21}$$

Theorem 11.1. *Let* $\hat{A}:\hat{\mathfrak{H}}\to\hat{\mathfrak{H}}^*$ *be in standard form and* $\hat{\mathfrak{I}}_P\subset\hat{\mathfrak{H}}$ *be isotropic for* $\hat{A}$. *Assume that* $\mathfrak{I}_{P1}$ *is closed in* $\mathfrak{H}$ *and*

$$\hat{\mathfrak{I}}_P\supset\mathfrak{N}_0^R. \tag{11.22}$$

Then $\hat{\mathfrak{I}}_P$ *is completely regular.*

Proof. Define $\hat{\mathfrak{H}}'=\mathfrak{I}_{P1}\oplus\mathfrak{I}_{P1}$ and

$$\hat{\mathfrak{I}}'_P=\hat{\mathfrak{H}}'\cap\hat{\mathfrak{I}}_P. \tag{11.23}$$

Observe that $\hat{\mathfrak{H}}'$ is a closed Hilbert subspace of $\hat{\mathfrak{H}}$.

Lemma 11.2. *Under the assumptions of Theorem* 11.1,

$$\hat{\mathfrak{I}}_P=\hat{\mathfrak{I}}'_P+\mathfrak{N}_0^R. \tag{11.24}$$

Proof. Let $\hat{u}=[u_1, u_2]\in\hat{\mathfrak{I}}_P$, then

$$u_2=u_2'+\bar{u}_2^{\perp}, \quad u_2'\in\mathfrak{I}_{P1} \quad \& \quad \bar{u}_2^{\perp}\in\mathfrak{I}_{P1}^{\perp}. \tag{11.25}$$

Therefore $\hat{u}=[u_1, u_2']+[0, u_2^{\perp}]$. Clearly $[u_1, u_2']\in\hat{\mathfrak{I}}'_P$ while $[0, u_2^{\perp}]\in\mathfrak{N}_0^R$. This shows $\hat{\mathfrak{I}}_P\subset\hat{\mathfrak{I}}'_P+\mathfrak{N}_0^R$. The converse of this relation is clear.

Let $\hat{A}':\hat{\mathfrak{H}}'\to(\hat{\mathfrak{H}}')^*$ be the restriction of $\hat{A}:\hat{\mathfrak{H}}\to(\hat{\mathfrak{H}})^*$ to $\hat{\mathfrak{H}}'$. Then $\hat{\mathfrak{I}}'_P\subset\hat{\mathfrak{H}}'$ is isotropic for $\hat{A}'$. Consider the problem with linear restrictions: given $U_1\in\mathfrak{I}_{P1}$ find $\hat{u}'=[u_1', u_2']\in\hat{\mathfrak{H}}'$ such that

$$\hat{u}'\in\hat{\mathfrak{I}}'_P \quad\text{while}\quad u_1'=U_1. \tag{11.26}$$

This problem satisfies existence, by virtue of the definition of $\mathfrak{I}_{P1}$. The unique solution of this problem (Corollary 9.2) will be denoted by $\sigma(U_1)\in\hat{\mathfrak{I}}'_P$. This defines a mapping $\sigma:\mathfrak{I}_{P1}\to\hat{\mathfrak{I}}'_P$. The space $\hat{\mathfrak{I}}'_P$ is completely regular with respect to $\hat{A}'$, by virtue of Corollary 9.1.

Given any $\hat{u}=[u_1, u_2]\in\hat{\mathfrak{H}}$, assume $\langle\hat{A}\hat{u}, \hat{v}\rangle=0$ for every $\hat{v}\in\hat{\mathfrak{I}}_P$. Taking $\hat{v}\in\mathfrak{N}_0^R$, it is seen that $(u_1, v_2^{\perp})=0$ for every $\bar{v}_2^{\perp}\in\mathfrak{I}_{P1}^{\perp}$, i.e., that $u_1\in\mathfrak{I}_{P1}$. Let (11.25) be satisfied, then $\hat{u}=[u_1, u_2']+[0, u_2^{\perp}]=\hat{u}'+\hat{u}_0$, where $\hat{u}'\in\hat{\mathfrak{I}}'_P$ and $\hat{u}_0\in\mathfrak{N}_0^R$. Observe, for every $\hat{v}=[v_1, v_2]\in\hat{\mathfrak{I}}_P$, one has

$$\langle\hat{A}\hat{u}_0, \hat{v}\rangle=(u_2^{\perp}, \bar{v}_1)=0, \tag{11.27}$$

because $v_1 \in \mathfrak{J}_{P1}$. Hence

$$\langle A\hat{u}, \hat{v}\rangle = \langle A\hat{u}', \hat{v}\rangle = \langle A\hat{u}', \hat{v}'\rangle = 0 \quad \forall\, \hat{v}' \in \hat{\mathfrak{J}}'_P \Rightarrow \hat{u}' \in \mathfrak{J}'_P \qquad (11.28)$$

and $\hat{u} = \hat{u}' + \hat{u}_0 \in \hat{\mathfrak{J}}_P$ by virtue of (11.24).

Corollary 11.1. *Under assumptions of Theorem* 11.1, *the mapping* $\sigma : \mathfrak{J}_{P1} \to \hat{\mathfrak{J}}'_P$ *is a topological isomorphism.*

Proof. This mapping is one-to-one (Corollary 9.2) and onto $\hat{\mathfrak{J}}'_P$. Given any $\hat{u} = [u_1, u_2] \in \hat{\mathfrak{J}}'_P$, $\sigma^{-1}(\hat{u}) = u_1$; thus $\sigma^{-1} : \hat{\mathfrak{J}}'_P \to \mathfrak{J}_{P1}$ is continuous (essentially a projection), one-to-one and onto the Hilbert space $\mathfrak{J}_{P1}$. Therefore, σ^{-1} is bicontinuous by Banach Theorem [103].

Theorem 11.2. *Let* $\hat{A} : \hat{\mathfrak{H}} \to \hat{\mathfrak{H}}^*$ *be in standard form. Assume that* $\hat{\mathfrak{J}}_P$ *is completely regular with respect to* $\hat{A}$ *and* $\mathfrak{J}_{P1}$ *is closed. Then a system* $\hat{\mathfrak{B}} = \{\hat{w}_1, \hat{w}_2, \ldots\} \subset \hat{\mathfrak{J}}_P$ *is T-complete for* $\hat{\mathfrak{J}}_P$ *with respect to* $\hat{A}$, *if and only if*

$$\operatorname{span} \mathfrak{B}_1 = \mathfrak{J}_{P1} \qquad (11.29a)$$

and simultaneously

$$\operatorname{span} \hat{\mathfrak{B}} \supset \hat{\mathfrak{N}}_0^R. \qquad (11.29b)$$

Proof. Using Propositions 11.1 and 11.2, it is seen that the T-completeness assumption implies (11.29). To prove the converse it is enough to show (11.2). Write $\hat{w}_\alpha = [w_{\alpha 1}, w_{\alpha 2}] \in \hat{\mathfrak{J}}_P$ and $\sigma(w_{\alpha 1}) = \hat{w}'_\alpha = [w_{\alpha 1}, w'_{\alpha 2}] \in \hat{\mathfrak{J}}'_P$, $\alpha = 1, 2, \ldots$. Let $\sigma(\mathfrak{B}_1) = \hat{\mathfrak{B}}' = \{\hat{w}'_1, \hat{w}'_2, \ldots\} \subset \hat{\mathfrak{J}}'_P$. Observe that $\hat{w}_\alpha - \hat{w}'_\alpha = [0, w_{\alpha 2} - w'_{\alpha 2}] \in \hat{\mathfrak{N}}_0^R \subset \operatorname{span} \hat{\mathfrak{B}}$. This shows that $\operatorname{span} \hat{\mathfrak{B}} \supset \operatorname{span} \hat{\mathfrak{B}}'$. Therefore

$$\operatorname{span} \hat{\mathfrak{B}} \supset \operatorname{span} \hat{\mathfrak{B}}' = \operatorname{span} \sigma(\mathfrak{B}_1) = \sigma(\operatorname{span} \mathfrak{B}_1) = \sigma(\mathfrak{J}_{P1}) = \hat{\mathfrak{J}}'_P.$$

This, together with (11.29b), shows that $\operatorname{span} \hat{\mathfrak{B}} \supset \hat{\mathfrak{J}}'_P + \hat{\mathfrak{N}}_0^R = \hat{\mathfrak{J}}_P$ and the proof of Theorem 11.2 is complete.

Theorem 11.3. *Let* $\hat{A} : \hat{\mathfrak{H}} \to \hat{\mathfrak{H}}^*$ *be in standard form and* $\hat{\mathfrak{J}}_P \subset \hat{\mathfrak{H}}$, *be completely regular for* $\hat{A}$. *Assume that*

(i) $\mu : D \to \hat{\mathfrak{H}}$ *is a symplectic homomorphism between* A, *D and* $\hat{A}$, $\hat{\mathfrak{H}}$,
(ii) $N_P \subset D$ *is a Hilbert space, and*
(iii) $N_P \cap N_A = \{0\}$; (11.30)

(iv) *the restriction* $\mu_r : N_P \to \hat{\mathfrak{J}}_P$ *of* μ *is continuous.*

Then $\mu_r : N_P \to \hat{\mathfrak{J}}_P$ *is a topological isomorphism.*

Proof. In view of (iii) and Lemma 11.1, $\mu_r : N_P \to \hat{\mathfrak{J}}_P$ is one-to-one. Also, $\hat{\mathfrak{J}}_P$ is closed in $\hat{\mathfrak{H}}$ by virtue of Corollary 10.1. Therefore, the mapping $\mu_r : N_P \to \hat{\mathfrak{J}}_P$ between the Hilbert spaces N_P and $\hat{\mathfrak{J}}_P$ is continuous, one-to-one and onto; thus, μ_r is a topological isomorphism by Banach Theorem [103].

Theorem 11.4. *Under the assumptions of Theorem* 11.3, *let* $\mathfrak{B} \subset N_P$ *and* $\hat{\mathfrak{B}} = \mu(\mathfrak{B})$. *Then the following assertions are equivalent:*

(a) $\mathfrak{B}$ *spans* N_P,
(b) $\mathfrak{B}$ *is T-complete for* I_P *with respect to* A;
(c) $\hat{\mathfrak{B}}$ *spans* $\hat{\mathfrak{J}}_P$; *and*
(d) $\hat{\mathfrak{B}}$ *is T-complete for* $\hat{\mathfrak{J}}_P$ *with respect to* $\hat{A}$.

Proof. (c) and (d) are equivalent by Proposition 11.1. (b) and (d) are equivalent because $\mu : D \to \hat{\mathfrak{H}}$ is a symplectic homomorphism between A, D and $\hat{A}$, $\hat{\mathfrak{H}}$. Finally, (a) and (c) are equivalent because $\mu_r : N_P \to \hat{\mathfrak{J}}_P$ is a topological isomorphism.

In view of the importance of these results, it is convenient to summarize them.

Theorem 11.5. *Let* $\mu : D \to \hat{\mathfrak{H}}$ *be a symplectic homomorphism from* D, A *onto* $\hat{\mathfrak{H}}$, $\hat{A}$, *where* $\hat{A}$ *is in standard form. Assume that:*
(I) $N_P \subset D$ *is isotropic for* A;
(II) $N_P \subset D$ *is a Hilbert space*;

(III) $N_P \cap N_A = \{0\}$; (11.31)

(IV) *the restriction* $\mu_r : N_P \to \hat{\mathfrak{J}}_P$ *of* μ *to* N_P *is continuous*;

(V) $\hat{\mathfrak{J}}_P \supset \hat{\mathfrak{N}}_0^R$, (11.32)

where $\hat{\mathfrak{N}}_0^R$ *is given by* (11.21); *and*
(VI) $\mathfrak{J}_{P1} \subset \mathfrak{H}$ *is closed in* $\mathfrak{H}$.

Then

(A) *the mapping* $\mu_r : N_P \to \hat{\mathfrak{J}}_P$ *is a topological isomorphism*;
(B) $I_P \subset D$ *and* $\hat{\mathfrak{J}}_P \subset \hat{\mathfrak{H}}$ *are completely regular for* A *and* $\hat{A}$, *respectively*;

(C) *given* $\mathfrak{B} \subset N_P$ *and* $\hat{\mathfrak{B}} \subset \hat{\mathfrak{I}}_P$ *such that* $\hat{\mathfrak{B}} = \mu(\mathfrak{B})$, *the following assertions are equivalent*:
(a) $\mathfrak{B}$ *spans* N_P,
(b) $\mathfrak{B}$ *is T-complete for* I_P *with respect to* A;
(c) $\hat{\mathfrak{B}}$ *spans* $\hat{\mathfrak{I}}_P$;
(d) $\hat{\mathfrak{B}}$ *is T-complete for* $\hat{\mathfrak{I}}_P$ *with respect to* $\hat{A}$; *and*
(e)
$$\text{span}\,\hat{\mathfrak{B}} \supset \hat{\mathfrak{N}}_0^R \tag{11.33a}$$

and simultaneously

$$\text{span}\,\mathfrak{B}_1 = \mathfrak{I}_{P1} \tag{11.33b}$$

Proof. This is clear from previous results.

In the remainder of this chapter a few applications are given of the theory that has been developed.

11.1 Laplace and Helmholtz equations

Let the linear space D be defined by (8.11) and $N_P \subset D$ be the linear manifold of functions $u \in D$ such that

$$\mathcal{L}u \equiv \Delta u + k^2 u = 0, \quad \text{in} \quad \Omega, \tag{11.34}$$

where $k = 0$ or 1.†

Proposition 11.3. *Define the linear space of functions D by means of equation* (8.11a). *Set* $\mathfrak{H} = H^\circ(\partial\Omega)$ *and* $\hat{\mathfrak{H}} = \mathfrak{H} \oplus \mathfrak{H}$. *Let* $\mu : D \to \hat{\mathfrak{H}}$ *be given by* (11.10), *while* $A : D \to D^*$ *is given by equation* (8.18). *The bilinear functional* $\hat{A} : \hat{\mathfrak{H}} \to \hat{\mathfrak{H}}^*$ *is taken in standard form so that equations* (11.16) *and* (11.17) *apply. Let*

$$N_P = \{u \in H^{3/2}(\Omega) \mid \mathcal{L}u = 0\} \tag{11.35}$$

Then $N_P \subset D$ *is a closed Hilbert subspace of* $H^{3/2}(\Omega)$ *and conclusions* (A) *to* (C) *of Theorem* 11.5 *hold.*

Proof. In view of continuity properties of elliptic operators [82], it is clear that $N_P \subset H^{3/2}(\Omega)$ is closed. In addition, let $u \in N_P \subset H^{3/2}(\Omega)$, then $\gamma_0 u \in H^1(\partial\Omega) \subset H^\circ(\partial\Omega)$ and $\gamma_1 u \in H^\circ(\partial\Omega)$, by virtue of general properties of the trace operators γ_0 and γ_1 [82]. This

† The case when k is purely imaginary can be treated along similar lines.

shows that $N_P \subset D$. The Hilbert space N_P will be denoted by $N^{3/2}(\Omega)$ and $N_h^{3/2}(\Omega)$, when $k=0$ and $k \neq 0$, respectively.

Let $u \in D$ satisfy (11.34). Then $\gamma_1 u \in H^{\circ}(\partial\Omega)$, which implies $u \in H^{3/2}(\partial\Omega)$; this shows that

$$N_P = \{u \in D \mid \mathcal{L}u = 0\}. \tag{11.36}$$

Applying Green's formula,

$$\int_{\Omega} \{v\mathcal{L}u - u\mathcal{L}v\} \, d\mathbf{x} = \int_{\partial\Omega} \left\{v\frac{\partial u}{\partial n} - u\frac{\partial v}{\partial n}\right\} d\mathbf{x}. \tag{11.37}$$

It can be seen that $N_P \subset D$ is isotropic for $A: D \to D^*$ as given by (8.18). Also

$$N_A = \{u \in D \mid \gamma_0 u = \gamma_1 u = 0\}. \tag{11.38}$$

Therefore, $u \in N_P \cap N_A$ if and only if $u \in D$ satisfies (11.34) and

$$u = \partial u/\partial n = 0, \quad \text{on} \quad \partial\Omega. \tag{11.39}$$

It is well known that any such function u is necessarily identically zero. This shows that

$$N_P \cap N_A = \{0\}. \tag{11.40}$$

This result is a particular case of a similar property that can be shown to hold for general formally symmetric differential equations. For elliptic equations, for example, it can be derived from general existence theorems [82, pp. 188–189]. Taking into account definition (11.10) of the mapping μ and continuity properties of the trace operators γ_0 and γ_1 [82], it is easy to see that the restriction mapping $\mu_r: N_P \to \hat{\mathfrak{J}}_P$ is continuous.

Finally, a general result [82, pp. 188–189] on the existence of solutions of elliptic differential equations shows that (11.22) is satisfied and also that $\mathfrak{J}_{P1} \subset \mathfrak{H} = H^{\circ}(\partial\Omega)$ is closed.

A more detailed discussion of the case $k=0$ may help to explain the meaning of these last assertions. The only harmonic functions $u \in H^{3/2}(\Omega)$ that satisfy

$$\gamma_1 u = \partial u/\partial n = 0, \quad \text{on} \quad \partial\Omega \tag{11.41}$$

are the constant functions. When $u \in H^{3/2}(\Omega)$ is a constant ($u = \lambda$), then

$$\mu(u) = [0, \lambda] \in \hat{\mathfrak{J}}_P, \tag{11.42}$$

where λ is a scalar. Also, the normal derivatives of harmonic functions $u \in H^{3/2}(\Omega)$ are restricted by the condition

$$\int_{\partial\Omega} \partial u/\partial n \, d\mathbf{x} = 0. \tag{11.43}$$

Using this equation and (11.10), it can be seen that

$$\mathfrak{I}_{P1} = \left\{ u_1 \in H^{\circ}(\partial\Omega) \,\middle|\, \int_{\partial\Omega} u_1 \, d\mathbf{x} = 0 \right\}, \tag{11.44}$$

by application of the general results given by Lions and Magenes [82, pp. 188–189]. Clearly, $\mathfrak{I}_{P1} \subset H^{\circ}(\partial\Omega) = \mathfrak{L}^2(\Omega)$ is a closed subspace of $\mathfrak{H} = H^{\circ}(\partial\Omega)$. Finally $\mathfrak{I}_{P1}^{\perp} \subset H^{\circ}(\partial\Omega)$ are the constant functions defined on $\partial\Omega$. Applying the second of equations (11.21), it is clear that relation (11.22) is satisfied by virtue of (11.42).

11.1.1 The source method

In the literature, procedures which are based on the use of a system of sources to represent any other field are frequently named under this heading. The foundations of such a method are given here.

Consider first the Laplace equation in two dimensions ($\Omega \subset \mathfrak{R}^2$). Let us revise some of the arguments presented in examples 4.7 and 4.8. Thus, Ω is a two-dimensional simply connected region and the complement of its closure (the exterior of Ω) is E (Fig. 11.1). Define $w_{\mathbf{y}}(\mathbf{x}) \in N_P(\mathbf{y} \in \mathsf{E})$ and $\mathfrak{B}_{\mathsf{E}} \subset N_P$ by means of equations (4.51) to (4.53). Let $u \in D$ and assume

$$\int_{\partial\Omega} \left\{ w_{\mathbf{y}} \frac{\partial u}{\partial n} - u \frac{\partial w_{\mathbf{y}}}{\partial n} \right\} d\mathbf{x} = \langle Au, w_{\mathbf{y}} \rangle = 0 \quad \forall\, \mathbf{y} \in \mathsf{E}. \tag{11.45}$$

The fact that $\gamma_0 u \in H^{\circ}(\partial\Omega)$ and $\gamma_1 u \in H^{\circ}(\partial\Omega)$ implies that $\gamma_0 u$ and $\gamma_1 u$ are continuous almost everywhere in $\partial\Omega$. When (11.45) is satisfied, well-known properties of surface and double distributions [95, pp. 160–175] grant that $\gamma_0 u = \gamma_0 v$ and $\gamma_1 u = \gamma_1 v$ where

$$v(\mathbf{y}) = \int_{\partial\Omega} \left\{ G(\mathbf{x}, \mathbf{y}) \frac{\partial u}{\partial n} - u \frac{\partial G(x, \mathbf{y})}{\partial n} \right\} d\mathbf{x}, \quad \mathbf{y} \in \Omega. \tag{11.46}$$

Such $v \in N_P$ and therefore $u \in I_P$. This shows that $\mathfrak{B}_{\mathsf{E}} \subset N_P$ is T-complete for I_P with respect to $A: D \to D^*$.

Define $\{w_0, w_1, \ldots\} \subset N_P = N^{3/2}(\Omega)$ by means of equations (4.54)

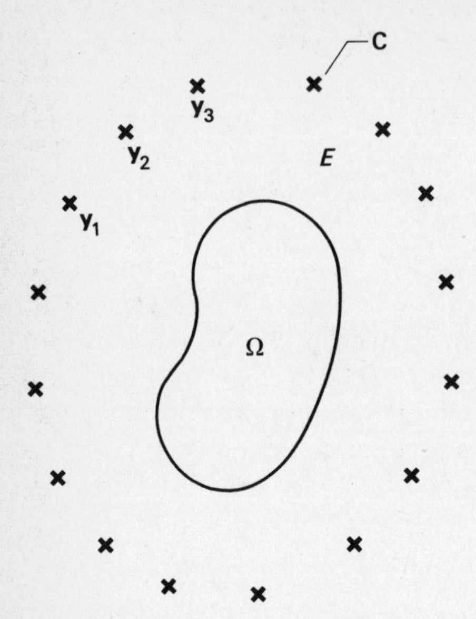

Fig. 11.1

and (4.56). Then the arguments presented in Example 4.8 show that $\mathfrak{B} \subset N^{3/2}(\Omega)$ is T-complete for $N^{3/2}(\Omega) = N_P$.

The following properties follow from Theorem 11.5 and Proposition 11.2.

Proposition 11.4. *If a system of sources given by equation* (4.51), *with singularities on a curve* $\mathfrak{C}$ (*Fig.* 11.1) *enclosing a simply connected and bounded region* Ω, *is supplemented with the constant function, the resulting system spans the space* $N^{3/2}(\Omega) \subset H^{3/2}(\Omega)$. *Here* $N^{3/2}(\Omega)$ *is the closed Hilbert subspace of* $H^{3/2}(\Omega)$, *which consists of harmonic functions belonging to* $H^{3/2}(\Omega)$. *The system of traces* $\{\gamma_0 w_0, \gamma_0 w_1, \ldots\}$ *spans* $H^{\circ}(\partial\Omega)$ *while the system* $\{\gamma_1 w_1, \gamma_1 w_2, \ldots\}$ *spans* $\{1\}^{\perp} \subset H^{\circ}(\partial\Omega)$. *Here* w_0 *is the constant function* 1, $\{1\}^{\perp} \subset H^{\circ}(\partial\Omega)$ *is the orthogonal complement of this function in* $H^{\circ}(\partial\Omega)$ *and the spans and orthogonal complements are taken in the* $H^{\circ}(\partial\Omega)$ *norm.*

The inclusion of the constant functions is only required if the curve $\mathfrak{C}$ happens to be anomalous, in the sense that zero is an eigenvalue for the exterior Dirichlet problem of the Laplace equation [96]. However, its inclusion in all cases is convenient, because such anomalous curves are difficult to recognize.

Proposition 11.4 also applies to the case when $\Omega \subset \mathfrak{R}^3$ is a three-dimensional region. However, the constant function is not needed because the solution of the exterior Dirichlet problem for functions with the required behaviour at infinity ($0(1/r)$), is always unique. In this case the curve $\mathfrak{C}$ has to be replaced by a closed surface which is the boundary of a simply connected region containing the closure of the region Ω in its interior.

The extension of these results to the Helmholtz equation is straightforward. The curve $\mathfrak{C}$ must be chosen so that the exterior Dirichlet problem satisfies uniqueness. Of course the constant function need not be included.

As mentioned previously, in the case of the Laplace equation in two dimensions it is possible to choose the curve $\mathfrak{C}$ arbitrarily if the system of sources is supplemented by the constant function in Ω. Similarly, for the Helmholtz equation the curve $\mathfrak{C}$ can be chosen arbitrarily if suitable additional functions are included in the system.

Notice, finally, that the results presented here are closely related to Kupradze's method of functional equations [104]. However, the point of view is different.

11.1.2 Separation of variables

A powerful procedure for obtaining systems of functions which are T-complete in arbitrary regions is the separation of variables.

By applying separation of variables in a unit circle for two-dimensional problems or in a unit sphere for three-dimensional problems, the systems of functions shown in Table 11.1 are obtained. $J_n(r)$ and $j_n(r)$ are Bessel functions and spherical Bessel functions, respectively, of the first class of order n [105–106].

TABLE 11.1 Complete systems for Laplace and Helmholtz equations

	Laplace equation	Helmholtz equation
Two dimensions	$r^n \cos n\theta$ $r^{n+1} \sin (n+1)\theta$	$J_n(r) \cos n\theta$ $J_{n+1}(r) \sin (n+1)\theta$
Three dimensions	$r^n Y^*_{nq}(\theta, \phi)$	$j_n(r) Y^*_{nq}(\theta, \phi)$

Here $n = 0, 1, 2, \ldots ;\ -n \leqslant q \leqslant n$.

$Y_{nq}(\theta, \phi)$ are normalized spherical harmonics defined by

$$Y_{nq}(\theta, \phi) = \left[\frac{2n+1}{4\pi}\frac{(n-q)!}{(n+q)!}\right]^{1/2} P_n^q(\cos\theta)e^{in\phi}. \qquad (11.47)$$

Here, $P_n^q(x)$ is the associated Legendre function, while

$$Y_{nq}(\theta, \phi) = (-1)^q Y_{n,-q}(\theta, \phi). \qquad (11.48)$$

In addition, $\{r, \theta\}$ and $\{r, \theta, \phi\}$ are polar and spherical coordinates, respectively. In each case, θ is the polar angle.

Proposition 11.5. *Let Ω be any bounded and simply connected region of $\Re^2$ or $\Re^3$. Then the systems of functions given in Table 11.1 span the closed Hilbert subspaces $N^{3/2}(\Omega)$ (Laplace equation) or $N_h^{3/2}(\Omega)$ (Helmholtz equation). The corresponding systems of traces γ_0 and γ_1 span $\Im_{P1}^c$ and $\Im_{P2}^c$, respectively, in the $\mathscr{L}^2(\partial\Omega) = H^o(\partial\Omega)$ norm. Here*

$$\Im_{P1} = \gamma_1 N_P, \quad \Im_{P2} = \gamma_0 N_P, \qquad (11.49)$$

where $N_P = N^{3/2}(\Omega)$ for the Laplace equation and $N_P = N_h^{3/2}(\Omega)$ for the Helmholtz equation. The superindex stands for the closure in $H^o(\partial\Omega) = \mathscr{L}^2(\partial\Omega)$.

Proof. In view of Proposition 11.3, it is enough to show that the systems given in Table 11.1 are T-complete for $I_P = N_P + N_A$ with respect to $A : D \to D^*$, when the latter is given by (8.18). This is immediate [25]. Indeed, Green's function for the Laplace equation in two dimensions [105, 106] is

$$G(\mathbf{x}, \mathbf{x}_0) = \frac{1}{2\pi}\left[\ln(1/r) + \sum_{n=1}^{\infty}\frac{1}{n}\left(\frac{r_0}{r}\right)^n \cos n(\phi - \phi_0)\right] \qquad (11.50)$$

and, in three dimensions, is

$$G(\mathbf{x}, \mathbf{x}_0) = \sum_{\ell=0}^{\infty}\sum_{n=-\ell}^{\ell}\frac{1}{2\ell+1}\frac{r_0^\ell}{r^{\ell+1}} Y_{\ell n}(\theta_0, \phi_0) Y_{\ell n}(\theta, \phi). \qquad (11.51)$$

For the Helmholtz equation in two dimensions,

$$G(\mathbf{x}, \mathbf{x}_0) = \frac{i}{4}\sum_{n=0}^{\infty} \epsilon_n J_n(kr_0) H_n^{(1)}(kr)\cos n(\phi - \phi_0), \qquad (11.52)$$

while in three dimensions

$$G(\mathbf{x}, \mathbf{x}_0) = ik\sum_{\ell=0}^{\infty}\sum_{n=-\ell}^{\ell} j_\ell(kr_0) h_\ell^{(1)}(kr) Y_{\ell n}^*(\theta_0, \phi_0) Y_{\ell n}(\theta, \phi) \qquad (11.53)$$

Here, it is assumed that $r > r_0$; r and r_0 have to be interchanged when $r < r_0$.

Let $\mathfrak{B} = \{w_1, w_2, \ldots\} \subset N_P$ be any of the systems given in Table 11.1. Recall that

$$\cos n(\phi - \phi_0) = \cos n\phi \cos n\phi_0 + \sin n\phi \sin n\phi_0. \tag{11.54}$$

Let $\rho_0 > 0$ be such that the ball of radius ρ_0 with centre at the origin of the coordinate system contains the closure of Ω in its interior (Fig. 11.2). Assume $|\mathbf{y}| > \rho_0$, then using (11.50) to (11.54) it is seen that

$$G(\mathbf{y}, \mathbf{x}) = \sum_{\alpha}^{\infty} a_\alpha(\mathbf{y}) w_\alpha(\mathbf{x}), \tag{11.55}$$

where $a_\alpha(\mathbf{y})$ are suitable functions of $\mathbf{y}$. By analyzing equations (11.50) to (11.53) it is easy to see that the series (11.55) is absolutely and uniformly convergent while $\mathbf{x}$ ranges over the closure of Ω. Therefore, for every $u \in D$, when $|\mathbf{y}| > \rho_0$, one has

$$\int_{\partial\Omega} \left\{ G(\mathbf{y}, \mathbf{x}) \frac{\partial u}{\partial n} - u(\mathbf{x}) \frac{\partial G(\mathbf{y}, \mathbf{x})}{\partial n_{\mathbf{x}}} \right\} d\mathbf{x} = \sum_{\alpha=1}^{\infty} a_\alpha(\mathbf{y}) \langle Au, w_\alpha \rangle \tag{11.56}$$

by virtue of (8.18). Consider the function

$$W(\mathbf{y}) = \int_{\partial\Omega} \left\{ G(\mathbf{y}, \mathbf{x}) \frac{\partial u}{\partial n} - u(\mathbf{x}) \frac{\partial G(\mathbf{y}, \mathbf{x})}{\partial n_{\mathbf{x}}} \right\} d\mathbf{x}, \tag{11.57}$$

which is defined for every $\mathbf{y} \in \mathsf{E}$. When $u \in D$ and

$$\langle Au, w_\alpha \rangle = 0 \quad \forall\, w_\alpha \in \mathfrak{B}, \tag{11.58}$$

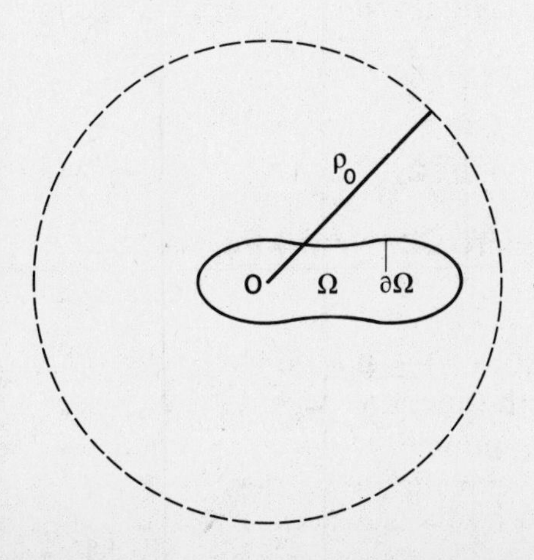

Fig. 11.2

equation (11.56) shows that $W(\mathbf{y})$ is identically zero outside a circle with radius ρ_0 and centre at the origin of the system of coordinates. By analytic continuation, which can be applied because $W(\mathbf{y})$ satisfies equation (11.34) at every $\mathbf{y} \in \mathsf{E}$, it is seen that $W(\mathbf{y})$ is identically zero in E. Hence,

$$\langle Au, w \rangle = 0 \quad \forall\, w \in \mathfrak{B}_{\mathsf{E}}. \tag{11.59}$$

Recall, however, that at the beginning of Section 11.1.1, it was shown that $\mathfrak{B}_{\mathsf{E}} \subset N_P$ is T-complete. Therefore,

$$\langle Au, w \rangle = 0 \quad \forall\, w \in \mathfrak{B} \Rightarrow u \in I_P, \tag{11.60}$$

because the premise in (11.60) implies (11.59) which, in turn, implies $u \in I_P$.

11.1.3 A system of plane waves

Let $\Omega \subset \Re^2$ be a bounded and simply connected region. Using the system $\{J_n(r) \cos n\theta;\ J_{n+1}(r) \sin n\theta \mid n = 0, 1, \ldots\} \subset N_h^{3/2}(\Omega)$ which has just been shown to span $N_h^{3/2}(\Omega)$ (Helmholtz equation) with respect to $A : D \to D^*$ (Equation 8.18), it is easy to exhibit a system whose elements are plane waves and which is also T-complete for $N_h^{3/2}(\Omega)$ [25].

Proposition 11.6. *Let $\{\mathbf{e}_1, \mathbf{e}_2, \ldots\}$ be a system of unit vectors in $\Re^2$, which is dense in the unit circle. Then the system*

$$\mathfrak{B} = \{e^{i\mathbf{e}_1 \cdot \mathbf{x}}, e^{i\mathbf{e}_2 \cdot \mathbf{x}}, \ldots\} \subset N_h^{3/2}(\Omega) \tag{11.61}$$

of plane waves which propagate in the directions $\mathbf{e}_1, \mathbf{e}_2, \ldots$, spans $N_h^{3/2}(\Omega)$.

Proof. Write

$$w_\alpha(\mathbf{x}) = e^{i\mathbf{e}_\alpha \cdot \mathbf{x}}, \quad \alpha = 1, 2, \ldots \tag{11.62}$$

and define for every real ξ of the interval $[-\pi, \pi)$ the functions

$$w_\xi(\mathbf{x}) = e^{ir \cos(\theta - \xi)}, \tag{11.63}$$

where r, θ are the polar coordinates of the position vector $\mathbf{x} \in \Re^2$. Notice that for every $\xi \in [-\pi, \pi)$ the function $w_\xi \in N_h^{3/2}(\Omega)$. Also, there is a sequence of angles (real numbers) $\{\psi_1, \psi_2, \ldots\}$ which is dense in the interval $[-\pi, \pi)$ and with the properties

(i) $\mathbf{e}_\alpha = \{\cos \psi_\alpha, \sin \psi_\alpha\}, \quad \alpha = 1, 2, \ldots$ (11.64)

and

(ii) $w_\alpha(\mathbf{x}) = w_{\psi\alpha}(\mathbf{x}).$ (11.65)

It is known that [107, p. 362]

$$J_n(r)e^{in\theta} = \frac{(-i)^n}{2\pi} \int_{-\pi}^{\pi} e^{in\xi + ir\cos(\theta-\xi)}\, d\xi. \quad (11.66)$$

Define

$$w_n'(r) = J_n(r)e^{in\theta}, \quad n = 0, \pm 1, \pm 2, \ldots. \quad (11.67)$$

Then the system $\mathfrak{B}' = \{w_0', w_1', \ldots\} \subset N_h^{3/2}(\Omega)$ is T-complete for the Helmholtz equation by virtue of Table 11.1. Using (11.63) and (11.66), one has

$$w_n'(\mathbf{x}) = \frac{(-i)^h}{2\pi} \int_{-\pi}^{\pi} e^{in\xi} w_\xi(\mathbf{x})\, d\xi. \quad (11.68)$$

Therefore, for every $u \in D$ one has

$$\langle Au, w_n' \rangle = \frac{(-i)^n}{2\pi} \int_{-\pi}^{\pi} e^{in\xi} \langle Au, w_\xi \rangle\, d\xi, \quad (11.69)$$

where the interchange of the integrals was possible because they, the integrands, are continuous on $\mathbf{x}$ and ξ. If $\langle Au, w \rangle = 0$ for every $w \in \mathfrak{B}$, then $\langle Au, w_\xi \rangle = 0$ for every $\xi = \psi_\alpha$, $\alpha = 1, 2, \ldots$, and the integral in (11.69) vanishes necessarily, since $\{\psi_1, \psi_2, \ldots\} \subset [-\pi, \pi)$ is dense in the interval $[-\pi, \pi)$ and the integrand in (11.69) is a continuous function. Hence

$$\langle Au, w \rangle = 0 \quad \forall\, w \in \mathfrak{B} \Rightarrow \langle Au, w' \rangle = 0 \quad \forall\, w' \in \mathfrak{B}' \Rightarrow u \in I_P. \quad (11.70)$$

This shows that $\mathfrak{B}$ is T-complete and therefore that $\mathfrak{B} \subset N_h^{3/2}(\Omega)$ spans $N_h^{3/2}(\Omega)$.

11.2 The biharmonic equation

The biharmonic equation is suitable for illustrating some peculiarities that occur when incorporating higher-order equations in the general framework developed in this book.

Define the linear space $D \subset D^e$ of functions $u \in D^e$ whose traces satisfy

$$u \in H^\circ(\partial\Omega), \quad \partial u/\partial n \in H^\circ(\partial\Omega), \quad \Delta u \in H^\circ(\partial\Omega), \quad \partial \Delta u/\partial n \in H^\circ(\partial\Omega). \quad (11.71)$$

Here, D^e is given by (8.4). More precisely,

$$D = \{u \in D^e \mid u \text{ satisfies } (11.71)\}. \tag{11.72}$$

The bilinear functional $A : D \to D^*$ will be (equation 4.41)

$$\langle Au, v\rangle = \int_{\partial\Omega} \left\{ v\frac{\partial \Delta u}{\partial n} - \Delta u \frac{\partial v}{\partial n} + \Delta v \frac{\partial u}{\partial n} - u\frac{\partial \Delta v}{\partial n} \right\} \mathrm{d}\mathbf{x}. \tag{11.73}$$

Define $\mathfrak{H} = H^{\circ}(\partial\Omega) \oplus H^{\circ}(\partial\Omega)$ and $\hat{\mathfrak{H}} = \mathfrak{H} \oplus \mathfrak{H}$. Motivated by the discussion presented in Example 4.6, let $\mu : D \to \hat{\mathfrak{H}}$ be given by

$$\mu(u) = [u_1, u_2], \quad u_1 \in \mathfrak{H} \quad \& \quad u_2 \in \mathfrak{H}, \tag{11.74}$$

where

$$u_1 = \left\{ \frac{\partial u}{\partial n}, \frac{\partial \Delta u}{\partial n} \right\} \quad \text{and} \quad u_2 = -\{\Delta u, u\} \tag{11.75}$$

Notice that u_1 and u_2 are pairs of functions defined in $\partial\Omega$; thus, u, Δu, $\partial\Delta u/\partial n$ and $\partial u/\partial n$ in (11.75) must be understood in the sense of traces. Let $\hat{A} : \hat{\mathfrak{H}} \to \hat{\mathfrak{H}}^*$ be in standard form, so that equations (11.16) and (11.17) apply. With these definitions it is clear that

$$N_{\mathrm{A}} = \left\{ u \in D \mid u = \frac{\partial u}{\partial n} = \Delta u = \frac{\partial \Delta u}{\partial n} = 0, \quad \text{on} \quad \partial\Omega \right\} \tag{11.76}$$

and

$$\langle Au, v\rangle = \langle \hat{A}\mu(u), \mu(v)\rangle. \tag{11.77}$$

Therefore $\mu : D \to \hat{\mathfrak{H}}$ is a symplectic homomorphism of A, D onto $\hat{A}, \hat{\mathfrak{H}}$.

The biharmonic equation is

$$\Delta^2 u = 0, \quad \text{in} \quad \Omega. \tag{11.78}$$

Thus, define

$$N_{\mathrm{P}} = \{u \in D \mid \text{equation (11.78) holds}\}. \tag{11.79}$$

In order to apply Theorem 11.5 to the biharmonic equation, it is necessary to supply N_{P} with a Hilbert space structure. First, a property that will be used in the sequel is established.

Proposition 11.7. *The only biharmonic functions $u \in N_{\mathrm{P}}$ that satisfy*

$$\frac{\partial u}{\partial n} = \frac{\partial \Delta u}{\partial n} = 0, \quad \text{on} \quad \partial\Omega \tag{11.80}$$

are the constant functions.

Proof. The system of boundary operators $\{\partial/\partial n, \partial\Delta/\partial n\}$ is a normal system [82, p. 113] for the biharmonic equation. Therefore, any $u \in N_P$ that satisfies equations (11.80) belongs to $\mathfrak{D}(\bar{\Omega})$ [82, p. 153]. Here, $\mathfrak{D}(\bar{\Omega})$ are the functions infinitely differentiable in the closure of Ω. Using this fact, one can apply the identity

$$\int_{\Omega} (\Delta u)^2 \, d\mathbf{x} = \int_{\Omega} u\Delta^2 u \, d\mathbf{x} - \int_{\partial\Omega} \left\{ u\frac{\partial \Delta u}{\partial n} - \Delta u \frac{\partial u}{\partial n} \right\} d\mathbf{x} = 0. \quad (11.81)$$

This shows that such a function u is harmonic. Once this has been shown, Proposition 11.7 is clear.

Define

$$N_b^{3/2}(\Omega) = \left\{ u \in N_P \,\middle|\, \frac{\partial \Delta u}{\partial n} = 0, \quad \text{on} \quad \partial\Omega \right\} \quad (11.82a)$$

and

$$N_b^{7/2}(\Omega) = \left\{ u \in N_P \,\middle|\, \int_{\partial\Omega} u \, d\mathbf{x} = 0 \quad \& \quad \frac{\partial u}{\partial n} = 0, \quad \text{on} \quad \partial\Omega \right\}. \quad (11.82b)$$

A simple result on the structure of $N_b^{3/2}(\Omega)$ will be needed. In what follows, $N^{3/2}(\Omega) \subset H^{3/2}(\Omega)$ is the closed subspace of harmonic functions introduced in Section 11.1.

Proposition 11.8. *Let $w_0 \in N_b^{3/2}(\Omega)$, be the unique solution of*

$$\Delta w_0 = 1, \quad \text{in} \quad \Omega \quad (11.83a)$$

subjected to the boundary condition

$$w_0 = 0, \quad \text{on} \quad \partial\Omega. \quad (11.83b)$$

Then

$$N_b^{3/2}(\Omega) = N^{3/2}(\Omega) + \{w_0\}. \quad (11.84)$$

Proof. Equation (11.84) is equivalent to $u = N_b^{3/2}(\Omega)$ if and only if

$$u = v + \lambda w_0, \quad v \in N^{3/2}(\Omega), \quad \lambda \in \mathfrak{R}. \quad (11.85)$$

Moreover, representation (11.85) is unique. Also, $w_0 \in \mathfrak{D}(\bar{\Omega})$.

Proposition 11.9. *The subspaces*

$$N_b^{3/2}(\Omega) \subset H^{3/2}(\Omega), \qquad N_b^{7/2}(\Omega) \subset H^{7/2}(\Omega) \quad (11.86)$$

are closed with respect to corresponding metrics.

Proof. $N_b^{3/2}(\Omega) \subset H^{3/2}(\Omega)$ is closed by virtue of Proposition 11.8, since $N^{3/2}(\Omega)$ is closed and $\{w_0\}$ is one-dimensional. The second of relations (11.86) can be established by observing that, when $u \in N_b^{7/2}(\Omega)$, by definition $\partial u/\partial n = 0$ while $\partial \Delta u/\partial n \in H^\circ(\partial\Omega)$. Hence, $u \in H^{7/2}(\Omega)$ by general existence theorems for elliptic equations [82, pp. 188–189]. Here, the fact, previously mentioned, that the system of boundary operators $\{\partial/\partial n, \partial\Delta/\partial n\}$ is normal for the biharmonic equation must be used. Now when $u \in H^{7/2}(\Omega)$, the traces $u \in H^3(\partial\Omega)$, $\partial u/\partial n \in H^2(\partial\Omega)$, $\Delta u \in H^1(\partial\Omega)$ and $\partial\Delta u/\partial n \in H^\circ(\partial\Omega)$. Thus all of these traces belong to $H^\circ(\partial\Omega)$. This shows that the space

$$N_b^{7/2}(\Omega) = \left\{ u \in H^{7/2}(\Omega) \mid \text{satisfy (11.78)}, \int_{\partial\Omega} u \, d\mathbf{x} = 0 \quad \& \quad \partial u/\partial n = 0, \quad \text{on} \quad \partial\Omega \right\}. \tag{11.87}$$

The continuity of the operators specifying the restrictions in (11.87) can be used to show that $N_b^{7/2}(\Omega)$ is a closed subspace of $H^{7/2}(\Omega)$; for example, the trace operator $\gamma_1 : H^{7/2}(\Omega) \to H^2(\partial\Omega) \subset H^\circ(\partial\Omega)$ is continuous.

Proposition 11.10. *Let $f_1 \in H^\circ(\partial\Omega)$ and $f_2 \in H^\circ(\partial\Omega)$. Then, the boundary value problem, "Find $u \in N_P$, such that*

$$\frac{\partial u}{\partial n} = f_1 \quad \textit{and} \quad \frac{\partial \Delta u}{\partial n} = f_2, \quad \textit{on} \quad \partial\Omega\text{''}, \tag{11.88}$$

possesses a solution if and only if

$$\int_{\partial\Omega} f_2 \, d\mathbf{x} = 0. \tag{11.89}$$

Such a solution is unique except by a constant function. Moreover,

$$u \in N_b^{3/2}(\Omega) \Leftrightarrow f_2 = 0 \tag{11.90a}$$

and

$$u \in N_b^{7/2}(\Omega) \Leftrightarrow f_1 = 0 \quad \textit{and} \quad \int_{\partial\Omega} u \, d\mathbf{x} = 0. \tag{11.90b}$$

Proof. This proposition can be established using Propositions 11.8 and 11.9, together with general existence theorems [82, pp. 188–189].

Proposition 11.11. *Every $u \in N_P$ can be written uniquely as*

$$u = u' + u'', \quad u' \in N_b^{3/2}(\Omega) \quad \& \quad u'' \in N_b^{7/2}(\Omega). \tag{11.91}$$

Proof. When $u \in N_P$, one necessarily has

$$\int_{\partial\Omega} \frac{\partial \Delta u}{\partial n} = 0. \tag{11.92}$$

Thus, given $u \in N_P$, let $u'' \in H^{7/2}(\Omega)$ be a solution of (11.78) such that

$$\frac{\partial u''}{\partial n} = 0 \quad \text{and} \quad \frac{\partial \Delta u''}{\partial n} = \frac{\partial \Delta u}{\partial n}, \quad \text{in} \quad \partial\Omega, \tag{11.93}$$

The existence of such u'' is granted by Proposition 11.10, since compatibility condition (11.89) is satisfied. Moreover, by addition of a constant one may require that $\int_{\partial\Omega} u'' \, d\mathbf{x} = 0$. Such $u'' \in N_b^{7/2}(\Omega)$ and $u' = u - u'' \in H^{3/2}(\Omega)$ satisfies (11.91). The uniqueness follows from the fact that

$$N_b^{3/2}(\Omega) \cap N_b^{7/2}(\Omega) = \{0\}, \tag{11.94}$$

which is implied by Proposition 11.7.

In view of Propositions 11.9 and 11.11, $N_P = N_b^{3/2}(\Omega) + N_b^{7/2}(\Omega)$ can be identified with the Hilbert space $N_b^{3/2}(\Omega) \oplus N_b^{7/2}(\Omega)$. This supplies $N_P = N_b^{3/2}(\Omega) + N_b^{7/2}(\Omega)$ with a Hilbert space structure.

Proposition 11.12. *Let the linear space of functions D be defined by means of equations (11.72) and (11.71), $\hat{\mathfrak{H}} = \mathfrak{H} \oplus \mathfrak{H}$ with $\mathfrak{H} = H^\circ(\partial\Omega) \oplus H^\circ(\partial\Omega)$, $\mu : D \to \hat{\mathfrak{H}}$ be given by (11.74) and (11.75) while $A : D \to D^*$ is the bilinear functional (11.73). The operator $\hat{A} : \hat{H} \to \hat{\mathfrak{H}}^*$ is defined by equations (11.16) and (11.17). In addition*

$$N_P = N_b^{3/2}(\Omega) + N_b^{7/2}(\Omega), \tag{11.95a}$$

$$\mathfrak{N}_0^R = \{[\{0, 0\}, \{0, \lambda\}] \mid \lambda \in H^\circ(\partial\Omega) \text{ is constant}\} \subset \hat{\mathfrak{H}} \tag{11.95b}$$

and

$$\mathfrak{J}_{P1} = H^\circ(\partial\Omega) \oplus \{1\}^\perp \subset \mathfrak{H}. \tag{11.95c}$$

Here $\{1\}^\perp \subset H^\circ(\partial\Omega)$ is the orthogonal complement, in $H^\circ(\partial\Omega)$, of the subspace of constant functions. Then, conclusions (A) to (C) of Theorem 11.5 hold.

Proof. Using the results of the previous discussion, assumptions (I) to (VI) are easy to verify. A few points deserve further attention. $N_P \subset D$ is isotropic for $A: D \to D^*$ because

$$\int_{\partial\Omega} \left\{ \Delta v \frac{\partial u}{\partial n} + v \frac{\partial \Delta u}{\partial n} \right\} d\mathbf{x} = \int_{\partial\Omega} \left\{ \Delta u \frac{\partial v}{\partial n} + u \frac{\partial \Delta v}{\partial n} \right\} d\mathbf{x} \tag{11.96}$$

whenever $u \in N_P$ and $v \in N_P$. In view of equation (11.76), $u \in N_A$ implies that (11.80) holds and simultaneously $u = 0$ on $\partial\Omega$. Hence, $u = 0$ in Ω, when $u \in N_P \cap N_A$ by virtue of Proposition 11.7. Let $u \in N_b^{3/2}(\Omega)$, then, using (11.74), (11.75) and the representation (11.85) of elements of $N_b^{3/2}(\Omega)$, it is seen that

$$\mu(u) = \left[\left\{ \frac{\partial u}{\partial n}, 0 \right\}, \{\lambda, u\} \right], \tag{11.97}$$

where $u \in H^1(\partial\Omega) \subset H^\circ(\partial\Omega)$, $\partial u/\partial n \in H^\circ(\partial\Omega)$ and λ is a constant function. The first two of these functions are the traces $\gamma_0 u$ and $\gamma_1 u$; they depend continuously on $u \in N^{3/2}(\Omega)$ by well-known continuity properties of elliptic differential equations [82]. To see that the constant function $\lambda \in H^\circ(\partial\Omega)$ also depends continuously on $u \in N_b^{3/2}(\Omega)$, let $e \in N_b^{3/2}(\Omega)$ be non-zero and satisfy $e \perp N^{3/2}(\Omega)$, where the orthogonality relation is in the sense of the $H^{3/2}(\Omega)$ inner product. Then

$$\lambda = (u, e)/(w_0, e) \tag{11.98}$$

by virtue of (11.85). In equation (11.98) the terms in parentheses refer to the inner product in $H^{3/2}(\Omega)$; therefore, it is clear that $\lambda \in H^\circ(\partial\Omega)$ depends continuously on $u \in H^{3/2}(\Omega)$. When $u \in N_b^{7/2}(\Omega)$, $\mu(u) \in [H_0(\partial\Omega)]^4$ depends continuously on u by well-known continuity properties of the trace operators [82; pp. 188–189]. This completes the proof of assumption (IV) of Theorem 11.5, because of Proposition 11.11.

Taking into account that the only restriction for existence of solution is (11.89), it can be seen that $\mathfrak{I}_{P1}$, as given by (11.95c), is consistent with the definition of N_P. Moreover, $\mathfrak{I}_{P1}$ is a closed subspace of $H^\circ(\partial\Omega) \oplus H^\circ(\partial\Omega)$ and assumption (IV) is fulfilled.

Using (11.95c) and (11.21), it is seen that equation (11.95b) is satisfied. According to Proposition 11.7, the only functions $u \in N_P$ which satisfy (11.80) are the functions which are constant in Ω. Let $1 \in N_P$ be the constant function 1, then

$$\mu(\mathbf{1}) = [\{0, 0\}, \{0, 1\}] \in \mathfrak{I}_P \subset \mathfrak{H} \tag{11.99}$$

by virtue of (11.74) and (11.75). Clearly, the space spanned by $\mu(1)$ is $\mathfrak{N}_0^R$. This shows assumption (V) of Theorem 11.5.

A systematic construction of complete systems for the biharmonic equation, starting from complete systems for the Laplace equation, is now easy to obtain [36].

Theorem 11.6. *Assume that* $\{\psi_1, \psi_2, \ldots\} \subset N^{3/2}(\Omega)$ *spans* $N^{3/2}(\Omega)$. *Let* $\{\phi_1, \phi_2, \ldots\} \subset N_b^{7/2}(\Omega)$ *be such that*

$$\Delta\phi_\alpha = \psi_\alpha, \quad \alpha = 1, 2, \ldots. \tag{11.100}$$

Then the system

$$\{\psi_1, \psi_2, \ldots\} \cup \{\phi_1, \phi_2, \ldots\} \subset N_b^{3/2}(\Omega) + N_b^{7/2}(\Omega) \tag{11.101}$$

spans $N_b^{3/2}(\Omega) + N_b^{7/2}(\Omega)$.

Proof. Here $N^{3/2}(\Omega) \subset H^{3/2}(\Omega)$ is the Hilbert space of harmonic functions (Section 11.1). To prove the theorem, we show that for every $u \in D$ one has that

$$\langle Au, \psi_\alpha \rangle = \langle Au, \phi_\alpha \rangle = 0 \quad \forall\, \alpha = 1, 2, \ldots \Rightarrow u \in I_P, \tag{11.102}$$

where $A : D \to D^*$ is given by (11.73), while $I_P = N_P + N_A$. Here, equations (11.76) and (11.79) hold. Let $u \in D$. Then

$$\langle Au, \psi_\alpha \rangle = \int_{\partial\Omega} \left\{ \psi_\alpha \frac{\partial \Delta u}{\partial n} - \Delta u \frac{\partial \psi_\alpha}{\partial n} \right\} d\mathbf{x} = 0 \quad \forall\, \alpha = 1, 2, \ldots. \tag{11.103}$$

Using the fact that the system $\{\psi_1, \psi_2, \ldots\} \subset N^{3/2}(\Omega)$ spans $N^{3/2}(\Omega)$, it is seen that condition (11.103) implies that there exists $w \in N^{3/2}(\Omega)$ such that

$$\Delta u = w \quad \text{and} \quad \frac{\partial \Delta u}{\partial n} = \frac{\partial w}{\partial n}, \quad \text{on} \quad \partial\Omega. \tag{11.104}$$

Take $v \in H^{3/2}(\Omega)$ with the property that $\Delta v = w$ in Ω, while $v = 0$ on $\partial\Omega$. This is possible by virtue of general existence theorems for boundary problems [82], since $w \in H^{3/2}(\Omega)$. Clearly $v \in D$, because

$$\gamma_0 v \in H^3(\partial\Omega) \subset H^\circ(\partial\Omega),$$
$$\gamma_1 v \in H^2(\partial\Omega) \subset H^\circ(\partial\Omega),$$
$$\gamma_0 \Delta v = \gamma_0 w \in H^\circ(\partial\Omega),$$
$$\gamma_1 \Delta v = \gamma_1 w \in H^\circ(\partial\Omega).$$

Also, $\Delta^2 v = \Delta w = 0$. Hence, $v \in N_P$. Observe that

$$\int_{\partial\Omega} \left\{ \phi_\alpha \frac{\partial \Delta u}{\partial n} - \Delta u \frac{\partial \phi_\alpha}{\partial n} \right\} = \int_{\partial\Omega} \left\{ \phi_\alpha \frac{\partial \Delta v}{\partial n} - \Delta v \frac{\partial \phi_\alpha}{\partial n} \right\} \mathbf{dx}$$

$$= -\int_{\partial\Omega} \left\{ \Delta\phi_\alpha \frac{\partial v}{\partial n} - v \frac{\partial \Delta\phi_\alpha}{\partial n} \right\} \mathbf{dx} = -\int_{\partial\Omega} \left\{ \psi_\alpha \frac{\partial v}{\partial n} - v \frac{\partial \psi_\alpha}{\partial n} \right\} \mathbf{dx}, \tag{11.105}$$

since $v \in N_P$. Therefore, the premise in (11.102) implies that

$$\int_{\partial\Omega} \left\{ \psi_\alpha \frac{\partial (u-v)}{\partial n} - (u-v) \frac{\partial \psi_\alpha}{\partial n} \right\} \mathbf{dx}$$

$$= \int_{\partial\Omega} \left\{ \psi_\alpha \frac{\partial u}{\partial n} - u \frac{\partial \psi_\alpha}{\partial n} + \phi_\alpha \frac{\partial \Delta v}{\partial n} - \Delta v \frac{\partial \phi_\alpha}{\partial n} \right\} \mathbf{dx}$$

$$= \int_{\partial\Omega} \left\{ \Delta\phi_\alpha \frac{\partial u}{\partial n} - u \frac{\partial \Delta\phi_\alpha}{\partial n} + \phi_\alpha \frac{\partial \Delta v}{\partial n} - \Delta v \frac{\partial \phi_\alpha}{\partial n} \right\} \mathbf{dx} = 0$$

$$\forall\, \alpha = 1, 2, \ldots. \tag{11.106}$$

Therefore, there exists $v' \in N^{3/2}(\Omega) \subset N_b^{3/2}(\Omega) \subset N_P \subset I_P$, such that

$$u - v = v' \quad \text{and} \quad \frac{\partial u}{\partial n} - \frac{\partial v}{\partial n} = \frac{\partial v'}{\partial n}, \quad \text{on} \quad \partial\Omega. \tag{11.107}$$

In addition, using (11.104) it is seen that

$$\Delta u - \Delta v = \Delta v' \quad \text{and} \quad \frac{\partial \Delta u}{\partial n} - \frac{\partial \Delta v}{\partial n} = \frac{\partial \Delta v'}{\partial n}, \quad \text{on} \quad \partial\Omega, \tag{11.108}$$

because v' is harmonic. Hence $u - v \in I_P$, which implies that $u \in I_P$ since $v \in N_P \subset I_P$. This establishes implication (11.102) and Theorem 11.6.

Corollary 11.2. *Assume that* $\{\psi_1, \psi_2, \ldots\} \subset N^{3/2}(\Omega)$ *spans* $N^{3/2}(\Omega)$. *Let* $\{p_1, p_2, \ldots\} \subset N^{3/2}(\Omega)$ *be such that for any fixed* $k = 1, 2, \ldots, n$ *(*n *is the dimension of the space* $\mathfrak{R}^n \supset \Omega$*),*

$$\frac{\partial p_\alpha}{\partial x_k} = \psi_\alpha, \quad \alpha = 1, 2, \ldots. \tag{11.109}$$

Then

(a) $\{\psi_1, \psi_2, \ldots\} \cup \{x_k p_1, x_k p_2, \ldots\} \subset N_b^{3/2}(\Omega) + N_b^{7/2}(\Omega)$ (11.110)

and

(b) $\operatorname{span}\{\psi_1, \psi_2, \ldots; x_k p_1, x_k p_2, \ldots\} = N_b^{3/2}(\Omega) + N_b^{7/2}(\Omega)$. (11.111)

Proof. In order to prove (a), it is enough to show that $x_k p_\alpha \in N_b^{3/2}(\Omega)+N_b^{7/2}(\Omega)$, for $\alpha=1,2,\ldots$, since $\psi_\alpha \in N^{3/2}(\Omega)\subset N_b^{3/2}(\Omega)$, for $\alpha=1,2,\ldots$. Now, $x_k p_\alpha \in H^{3/2}(\Omega)$, because $x_k \in \mathfrak{D}(\bar{\Omega})$ while $\psi_\alpha \in H^{3/2}(\Omega)$. In addition,

$$\Delta(x_k p_\alpha)=2\frac{\partial p_\alpha}{\partial x_k}=2\psi_\alpha. \tag{11.112}$$

This shows that $x_k p_\alpha$ is biharmonic. This, together with the fact that the traces $\gamma_0(x_k p_\alpha)\in H^1(\partial\Omega)$, $\gamma_1(x_k p_\alpha)\in H^\circ(\partial\Omega)$, $\gamma_0(\Delta x_k p_\alpha)=2\gamma_0(\psi_\alpha)\in H^1(\partial\Omega)$ and $\gamma_1(\Delta x_k p_\alpha)=2\gamma_1(\psi_\alpha)\in H^\circ(\partial\Omega)$, shows that $x_k p_\alpha \in N_b^{3/2}(\Omega)+N_b^{7/2}(\Omega)$, for $\alpha=1,2,\ldots$, and the proof of (a) is complete. Once this has been shown, application of Theorem 11.6 yields the desired result.

As an application of Corollary 11.2, let $\{\psi_1,\psi_2,\ldots\}\subset N^{3/2}(\Omega)$ be the system of functions given in Table 11.1, for the Laplace equation in two dimensions. Denoting by x and y the Cartesian coordinates in the plane, one has

$$\frac{\partial}{\partial x}[r^n e^{in\theta}]=nr^{n-1}e^{i(n-1)\theta}, \quad n=1,2,\ldots. \tag{11.113}$$

Therefore

$$\frac{\partial}{\partial x}(r^n\cos n\theta)=nr^{n-1}\cos(n-1)\theta,$$

$$\frac{\partial}{\partial x}(r^n\sin n\theta)=nr^{n-1}\sin(n-1)\theta. \tag{11.114}$$

Define

$$\psi_\alpha=r^{(\alpha-1)/2}\cos(\alpha-1)\theta/2, \quad \alpha=1,3,5,\ldots, \tag{11.115a}$$

$$\psi_\alpha=r^{\alpha/2}\cos\alpha\theta/2, \quad \alpha=2,4,\ldots. \tag{11.115b}$$

Take

$$p_\alpha=\psi_{\alpha+2}. \tag{11.116}$$

Corollary 11.2 shows that the system of biharmonic functions

$$r^n\cos n\theta, \quad r^{n+1}\sin(n+1)\theta, \quad n=0,1,\ldots, \tag{11.117a}$$

together with

$$r^{n+1}\cos\theta\cos n\theta, \quad r^{n+1}\cos\theta\sin n\theta, \quad n=1,2,\ldots \tag{11.117b}$$

spans $N_b^{3/2}(\Omega)+N_b^{7/2}(\Omega)$.

11.3 General elliptic equations

It is of interest, because it exhibits the generality of the procedures presented in this book, to cast the most general (formally symmetric) elliptic equation in the framework developed in this chapter. Of course, as has already been mentioned, there are many alternative manners of carrying out the immersion of the problem in a Hilbert space structure. Here, it is not intended to carry out an exhaustive analysis of such alternatives.

Using the notation introduced in Section 6.1, in order to formulate general existence theorems for elliptic equations, the following assumptions are needed [82]:

(i) *Ω is a bounded open set in $\mathfrak{R}^n$ with boundary $\partial\Omega$, an $(n-1)$-dimensional infinitely differentiable variety, Ω being locally on one side of $\partial\Omega$;*
(ii) *the operator $\mathscr{L}$ is defined by*

$$\mathscr{L}u = \sum_{|p|,\,|q|\leq m} (-1)^{|p|} D^p(a_{qp}(\mathbf{x})D^q u), \tag{11.118}$$

with $a_{qp} \in \mathfrak{D}(\bar{\Omega})$ and $a_{qp} = a_{pq}$.
(iii) *the operators C_j are defined by*

$$C_j u = \sum_{|h|\leq m_j} c_{jh}(\mathbf{x}) D^h u, \tag{11.119}$$

with $c_{jh} \in \mathfrak{D}(\partial\Omega)$, $0 \leq m_j \leq 2m-1$, the system $\{C_j\}_{j=0}^{m-1}$ being normal on $\partial\Omega$ and covering $\mathscr{L}$ on $\partial\Omega$.

The reader is referred to the treatise by Lions and Magenes [82] for detailed conditions under which a system of operators $\{C_j\}_{j=0}^{m-1}$ covers $\mathscr{L}$. Among the systems of operators that satisfy assumption (iii) for every properly elliptic operator $\mathscr{L}$, we will consider the system

$$C_j = \gamma_{m+j}, \quad j = 0, \ldots, m-1, \tag{11.120}$$

where γ_j are the trace operators $\gamma_j = \partial^j/\partial n^j$, with $\mathbf{n}$ normal to $\partial\Omega$ oriented towards the exterior of Ω.

In what follows it will be assumed that the system $\{C_j\}_{j=0}^{m-1}$ satisfies condition (iii) and in addition that the orders m_j of C_j $(j = 0, \ldots, m)$ are

$$m_j = m + j. \tag{11.121}$$

Thus, a system that satisfies these conditions is given by (11.20).

Given $\{C_j\}_{j=0}^{m-1}$, choose (non-uniquely) a system of operators $\{\mathfrak{S}_j\}_{j=0}^{m-1}$, of orders

$$\mu_j = j, \quad j = 0, \ldots, m-1, \tag{11.122}$$

so that the system $\{C_0, \ldots, C_{m-1}, \mathfrak{S}_0, \ldots, \mathfrak{S}_{m-1}\}$ is a Dirichlet system of order $2m$ on $\partial\Omega$.

Define the linear subspace $D \subset D^e$ of functions $u \in D^e$ (D^e given by (8.4)) whose traces satisfy

$$C_j u \in H^\circ(\partial\Omega) \quad \text{and} \quad \mathfrak{S}_j u \in H^\circ(\partial\Omega), \quad j = 0, \ldots, m-1. \tag{11.123}$$

The bilinear functional $A: D \to D^*$ will be

$$\langle Au, v\rangle = \sum_{j=0}^{m-1} \int_{\partial\Omega} \{\mathfrak{S}_j u\, C_j v - \mathfrak{S}_j v\, C_j u\}\, d\mathbf{x}. \tag{11.124}$$

Define $\mathfrak{H} = \underbrace{H^\circ(\partial\Omega) \oplus \cdots \oplus H^\circ(\partial\Omega)}_{m \text{ factors}}$ and $\hat{\mathfrak{H}} = \mathfrak{H} \oplus \mathfrak{H}$.

Let $\mu: D \to \hat{\mathfrak{H}}$ be $\mu(u) = [u_1, u_2]$, where

$$u_1 = \{C_0 u, \ldots, C_{m-1} u\}, \quad u_2 = \{\mathfrak{S}_0 u, \ldots, \mathfrak{S}_{m-1} u\}. \tag{11.125}$$

Clearly, $\mu: D \to \hat{\mathfrak{H}}$ is onto and

$$\langle Au, v\rangle = \langle \hat{A}\mu(u), \mu(v)\rangle, \tag{11.126}$$

where $\hat{A}: \hat{\mathfrak{H}} \to \hat{\mathfrak{H}}^*$ is given by (11.16). Thus, μ is a symplectic homomorphism of A, D onto $\hat{A}$, $\hat{\mathfrak{H}}$. Let

$$N_P = \{u \in D \mid \mathcal{L}u = 0 \quad \text{in} \quad \Omega\}. \tag{11.127}$$

Define

$$N_P^R = \{u \in N_P \mid \mathfrak{S}_j u = 0, \quad j = 0, \ldots, m-1\}. \tag{11.128}$$

Then, every $u \in N_P$ can be written uniquely as

$$u = u^R + \sum_{j=0}^{m-1} u^{(j)} \tag{11.129}$$

where $u^R \in N_P^R$ and, for every $k = 0, \ldots, m-1$,

$$C_j u^{(k)} = 0, \quad j \neq k, \tag{11.130a}$$

while

$$C_k u^{(k)} = C_k u. \tag{11.130b}$$

In addition, it is required that $u_2^k \in \mathfrak{I}_{P1} = \mathfrak{I}_{P1} \subset \mathfrak{H}$. Here, the superindex c stands for the closure in $\mathfrak{H}$ and it is assumed that $\mu(u^{(k)})=[u_1^k, u_2^k]\in\hat{\mathfrak{H}}=\mathfrak{H}\oplus\mathfrak{H}$. To supply N_P^R with a Hilbert-space structure is easy, because it is finite-dimensional [82]. Given the operators $\mathcal{L}$, $\{C_j\}_{j=0}^{m-1}$, $\{\mathfrak{S}_j\}_{j=0}^{m-1}$, define the linear subspace $N_{\mathcal{L}C\mathfrak{S}}^{m+k+(1/2)}(\Omega)$ for every $k=0,\ldots,m-1$, as the collection of elements $u\in N_P$ such that $C_ju=0$ when $j\neq k$ while $u_2\in\mathfrak{J}_{P1}^c$. It is easy to see that $N_{\mathcal{L}C\mathfrak{S}}^{m+k+(1/2)}(\Omega)\subset H^{m+k+(1/2)}(\Omega)$ is a closed subspace of $H^{m+k+(1/2)}(\Omega)$. Hence, N_P becomes a Hilbert space when it is supplied with the structure of $N_P^R\oplus N_{\mathcal{L}C\mathfrak{S}}^{m+(1/2)}(\Omega)\oplus\cdots\oplus N_{\mathcal{L}C\mathfrak{S}}^{2m-(1/2)}(\Omega)$. The verification of assumptions (I) to (VI) is now straightforward (Theorem 11.5).

12 Solution of boundary value problems

The algebraic theory of boundary value problems that has been developed throughout this book is here applied to formulate general algorithms for constructing solutions of boundary value problems. Two alternative algorithms will be presented.

Given a regular subspace $I \subset D$ and a commutative subspace $N_P \subset D$, the abstract boundary value problem requires finding $u \in D$ such that

$$u - U \in N_P \quad \text{while} \quad u - V \in I, \tag{12.1}$$

where $U \in D$ and $V \in D$ are data of the problem.

By considering $u' = u - U$ this problem can be transformed into one for which $U = 0$. Thus, attention will be restricted to the case $U = 0$, i.e., to the problem of finding $u \in D$ such that

$$u \in N_P \quad \text{while} \quad u - V \in I. \tag{12.2}$$

In order to apply the results of Chapters 10 and 11 it will be assumed that there is a mapping $\mu : D \to \hat{\mathfrak{H}} = \mathfrak{H} \oplus \mathfrak{H}$ where $\mathfrak{H}$ is a Hilbert space, such that for every $u \in D$ one has $u \overset{\mu}{\Rightarrow} \hat{u} = [u_1, u_2] \in \hat{\mathfrak{H}}$ with the property that

$$u \in I \Leftrightarrow u_1 = 0. \tag{12.3}$$

It will be assumed that $\mu : D \to \hat{\mathfrak{H}}$ is a symplectic homomorphism between D, A and $\hat{A}$, $\hat{\mathfrak{H}}$, where $\hat{A} : \hat{\mathfrak{H}} \to \hat{\mathfrak{H}}^*$ is in standard form. Moreover, conclusions (A) to (C) of Theorem 11.5 hold, but not necessarily the assumptions. Thus, N_P and $\hat{\mathfrak{I}}_P$ are Hilbert spaces. Also, (11.31) holds. The linear subspace $I_P = N_P + N_A$ is completely regular.

In view of (12.3), condition (12.2) can be written as

$$u \in N_P \quad \text{while} \quad u_1 = V_1. \tag{12.4}$$

Two alternative approaches for constructing a solution u will be given. Both are based on the use of T-complete systems. According to Proposition 11.2, when a system $\mathfrak{B} = \{\hat{w}_1, \hat{w}_2, \ldots\} \subset \hat{\mathfrak{I}}_P$ is

T-complete for $\hat{\mathfrak{I}}_P$, then

$$\text{span}\,\mathfrak{B}_1 = \mathfrak{I}^c_{P1} \quad\text{and}\quad \text{span}\,\mathfrak{B}_2 = \mathfrak{I}^c_{P2}. \tag{12.5}$$

This property will be basic for our constructions. In one of them, approximations of the boundary data are constructed and in the other the unknown complementary boundary values are approximated.

12.1 Least squares on the prescribed boundary values

Given $V_1 \in \mathfrak{H}$, the abstract boundary value problem requires finding $u \in N_P$ which satisfies (12.4). Clearly, a solution exists if and only if $V_1 \in \mathfrak{I}_{P1}$.

Proposition 12.1. *Let $\hat{\mathfrak{I}}_P \subset \hat{\mathfrak{H}}$ be completely regular for $\hat{A}:\hat{\mathfrak{H}} \to \hat{\mathfrak{H}}^*$. For every $V_1 \in \mathfrak{I}_{P1}$, the abstract boundary problem possesses a unique solution such that*

$$u_2 \in \mathfrak{I}_{P2} \cap \mathfrak{I}^c_{P1} \tag{12.6}$$

Proof. When $V_1 \in \mathfrak{I}_{P1}$, a solution $u \in N_P$ necessarily exists. Let $\hat{u} = [u_1, u_2] = \mu(u) \in \hat{\mathfrak{I}}_P$. When $\hat{\mathfrak{I}}_P$ is completely regular for $\hat{A}$, $\mathfrak{I}_{P2} \supset \mathfrak{I}^\perp_{P1}$. Let $u'_2 \in \mathfrak{I}^c_{P1}$ be the projection of u_2 on $\mathfrak{I}^c_{P1}$, then $\hat{u}' = [u_1, u'_2] \in \hat{\mathfrak{I}}_P$ is also a boundary solution of the problem and $u'_2 \in \mathfrak{I}_{P2} \cap \mathfrak{I}^c_{P1}$. This shows that $\mu^{-1}(\hat{u}') \in N_P$ is a solution which satisfies (12.6). In addition, a solution satisfying (12.6) is unique because when $u_1 = 0$, $u_2 \in \mathfrak{I}^\perp_{P1}$. Thus, when $u_1 = 0$ and condition (12.6) hold simultaneously, one has

$$u_2 \in \mathfrak{I}^c_{P1} \cap \mathfrak{I}^\perp_{P1} = \{0\}.$$

Therefore, $\mu(u) = [0, 0] \Rightarrow u = 0$.

Let

$$N^u_P = \{u \in N_P \mid u_2 \in \mathfrak{I}_{P2} \cap \mathfrak{I}^c_{P1}\}. \tag{12.7}$$

When $\hat{\mathfrak{I}}_P$ is completely regular, it can be seen that N^u_P is closed. In view of Proposition 12.1, there is a mapping $\sigma: \mathfrak{I}_{P1} \to N^u_P$ with the property that for every $V_1 \in \mathfrak{I}_{P1}$, $\sigma(V_1) \in N^u_P$ is the unique solution of the abstract boundary value problem which satisfies (12.6). The space $N_P \supset N^u_P$ is assumed to be a Hilbert space.

However, $\mathfrak{I}_{P1} \subset \mathfrak{H}$ is not, in general, a closed Hilbert space. Thus, $\sigma: \mathfrak{I}_{P1} \to N^u_P$ is not a bounded mapping. Let $\|\ \|_u$ be a norm defined on N_P, in which σ is bounded (below and above). By continuity it is then possible to extend σ to a mapping defined on $\mathfrak{I}_{P1}^c$ with values on N_P^c, where the latter is the closure in the norm $\| \|_u$. Thus, the function σ has been extended to a mapping $\sigma^e : \mathfrak{I}_{P1}^c \to (N_P^u)^c$. Moreover, σ^e is topological isomorphism; therefore

$$\sigma^e(\mathfrak{I}_{P1}^c) = (N_P^u)^c. \tag{12.8}$$

Let $N_{P0}^R \subset N_P$ be such that

$$\mu(N_{P0}^R) = \mathfrak{N}_0^R \tag{12.9}$$

where $\mathfrak{N}_0^R$ is given by (11.21). Now, $\mathfrak{N}_0^R$ is closed and, therefore, so is N_{P0}^R. Also, it can be seen that

$$(N_P)^c = (N_P^u)^c + N_{P0}^R. \tag{12.10}$$

Let $D^e = D + (N_P)^c$, then one can formulate a generalized version of the abstract boundary value problem:

"Given $V_1 \in \mathfrak{I}_{P1}^c$, $u \in D^e$ is said to be a solution of the generalized version of the abstract boundary value problem if

$$u = \sigma^e(V_1) + u^o, \tag{12.11}$$

with $u^o \in N_P^R$."

Existence of a solution to this problem is granted when $V_1 \in \mathfrak{I}_{P1}^c$.

Proposition 12.2. *Let $\mathfrak{B} = \mathfrak{B}_0 \cup \{w_1, w_2, \ldots\} \subset N_P$ be a T-complete system for N_P with respect to $A : D \to D^*$. Assume*

$$\text{span } \mathfrak{B}_0 = \mathfrak{N}_0^R \tag{12.12a}$$

and

$$\langle A\hat{w}_\alpha, \hat{w} \rangle = 0 \quad \forall\, \hat{w} \in \mathfrak{B}_0 \quad \& \quad \alpha = 1, 2, \ldots. \tag{12.12b}$$

Consider the approximating sequence

$$u^N = \sum_{\alpha=1}^{N} a_\alpha^N w_\alpha, \quad N = 1, 2, \ldots, \tag{12.13}$$

where the coefficients a_α^N ($\alpha = 1, 2, \ldots, N$) satisfy

$$\sum_{\alpha=1}^{N} a_\alpha^N (w_{\alpha 1}, w_{\beta 1}) = (V_1, w_{\beta 1}), \quad \beta = 1, \ldots, N. \tag{12.14}$$

Then, $u^N \to u$, where the convergence is with respect to the norm $\| \|_u$.

Proof. Conditions (12.12) imply that $w_\alpha \in N_P^u$, $\alpha = 1, 2, \ldots$. In addition, conditions (12.14) imply that $u_i^N \in \mathfrak{I}_{P1}$ is the projection of V_1 on the N-dimensional space spanned by $\{w_{11}, \ldots, w_{N1}\} \subset \mathfrak{I}_{P1}$. Therefore $u_1^N \to V_1 \in \mathfrak{I}_{P1}^c$, by virtue of the first of equations (11.5).

Example 12.1. For the Laplace equation, with the definitions adopted in Section 11.1,

$$\mathfrak{I}_{P2} \cap \mathfrak{I}_{P1}^c = \{1\}^\perp. \tag{12.15}$$

In this case, $\mathfrak{I}_{P1} \subset H^\circ(\partial\Omega)$, is closed; indeed

$$\mathfrak{I}_{P1} = \{1\}^\perp = \mathfrak{I}_{P1}^c. \tag{12.16}$$

Also,

$$N_P^u = \left\{ u \in N^{3/2}(\Omega) \,\middle|\, \int_{\partial\Omega} u \, d\mathbf{x} = 0 \right\}. \tag{12.17}$$

Observe that N_P^u is *not* the orthogonal complement of the constant function 1, in the $H^{3/2}(\Omega)$ inner product. In this case, given any $V_1 \in \{1\}^\perp \subset H^\circ(\partial\Omega)$, $\sigma(V_1) \in N^{3/2}(\Omega)$ is the unique solution of Neuman's problem which satisfies

$$\int_{\partial\Omega} u \, d\mathbf{x} = 0. \tag{12.18}$$

The mapping $\sigma : \mathfrak{I}_{P1} \to N_P^u \subset H^{3/2}(\Omega)$ is bicontinuous when the norms in $\mathfrak{I}_{P1}$ and N_P^u are those associated with $H^\circ(\partial\Omega)$ and $H^{3/2}(\Omega)$, respectively. The fact that $\mathfrak{I}_{P1}$ and $N^{3/2}(\Omega) = N_P$ are closed implies that the extended mapping σ^e is the same as σ and that the generalized version of the abstract boundary value problem is the same as the original version of that problem.

Example 12.2. Let us consider again the Laplace equation. In Section 11.1, modify the definition of $\mu : D \to \mathfrak{H}$ taking

$$\mu(u) = [\gamma_0 u, \gamma_1 u]. \tag{12.19}$$

Then

$$\mathfrak{I}_{P1} = H^1(\partial\Omega) \subset H^\circ(\partial\Omega) \tag{12.20}$$

$$\mathfrak{I}_{P2} \cap \mathfrak{I}_{P1}^c = \mathfrak{I}_{P2} = \{\mathbf{1}\}^\perp. \tag{12.21}$$

All orthogonal complements are taken in $H^\circ(\partial\Omega)$. In view of (12.20), $\mathfrak{I}_{P1}$ is dense in $\mathfrak{H} = H^\circ(\partial\Omega)$, but not closed. Also,

$$N_P^u = N_P = N^{3/2}(\Omega), \tag{12.22}$$

since the orthogonal complement of $\mathfrak{I}_{P1}$ is $\{0\}$. The norm used in $\mathfrak{I}_{P1}=H^1(\partial\Omega)$ is that associated with $H^\circ(\partial\Omega)$. Therefore, $\sigma:\mathfrak{I}_{P1}\to N_P^u=N^{3/2}(\Omega)$ is not continuous in the $H^{3/2}(\Omega)$ norm used in N_P^u. However, let $\|\ \|_u$ be the norm associated with $H^{1/2}(\Omega)$. In this norm the mapping σ of $\mathfrak{I}_{P1}\subset H^\circ(\partial\Omega)$ into $N^{3/2}(\Omega)\subset H^{1/2}(\Omega)$ is bicontinuous (bounded above and below). The extension of σ, $\sigma^e:\mathfrak{I}_{P1}^c\to H^{1/2}(\Omega)$, is a topological isomorphism. Here $\mathfrak{I}_{P1}^c=H^\circ(\partial\Omega)$. The algorithm introduced in Proposition 12.2 yields an approximation which converges in $H^{1/2}(\Omega)$. The generalized version of the boundary problem admits any function $V_1\in H^\circ(\partial\Omega)$.

12.2 Least squares on the complementary boundary values

This approach can be used to obtain a boundary solution (Definition 9.3). The corresponding problem is:

"Given $V_1\in\mathfrak{H}$, find $\hat{u}=[u_1, u_2]\in\hat{\mathfrak{I}}_P=\mathfrak{N}_P\subset\hat{\mathfrak{H}}$,

$$\text{such that } u_1=V_1.\text{"} \tag{12.23}$$

Equivalently: "Given $V_1\in\mathfrak{H}$, find $u_2\in\mathfrak{H}$ such that $[V_1, u_2]\in\hat{\mathfrak{I}}_P$."

Proposition 12.3. *Assume that $\hat{\mathfrak{I}}_P$ is completely regular and*

$$\mathfrak{B}=\{\hat{w}_1, w_2, \ldots\}\subset\hat{\mathfrak{I}}_P\subset\hat{\mathfrak{H}} \tag{12.24}$$

is T-complete for $\hat{\mathfrak{I}}_P$. Then, for every $V_1\in\mathfrak{H}$, the abstract boundary value problem possesses a boundary solution (not necessarily unique) if and only if the sequence u_2^N ($N=1,\ldots$), defined by

$$u_2^N=\sum_{\alpha}^{N} a_\alpha^N \bar{w}_{\alpha_1} \tag{12.25}$$

with

$$\sum_{\alpha=1}^{N} a_\alpha^N(\bar{w}_{\alpha1}, \bar{w}_{\beta1})=(w_{\beta2}, \bar{V}_1),\quad \beta=1,\ldots,N, \tag{12.26}$$

converges.

When a boundary solution exists, any boundary solution is given by

$$\hat{u}=\hat{u}'+\hat{q}, \tag{12.27a}$$

with

$$\hat{u}=[V_1, u_2'],\quad \hat{q}\in\mathfrak{N}_P^R \tag{12.27b}$$

and

$$u_2' = \lim_{N\to\infty} u_2^N. \tag{12.28}$$

Proof. Clearly, there is a boundary solution if and only if $V_1 \in \mathfrak{I}_{P1}$. Assume $V_1 \in \mathfrak{I}_{P1}$ and $[V_1, u_2] \in \mathfrak{N}_P$. Write $u_2 = u_2' + q_2$, with $u_2 \in \mathfrak{I}_{P1}^c$, $q_2 \in \mathfrak{I}_{P1}^\perp$. Then (12.27) are satisfied, with $\hat{q} = [0, q_2] \in \mathfrak{N}_P^R$. Notice that u_2^N is defined uniquely by (12.25) and (12.26), even when the matrix $(\bar{w}_{\alpha 1}, \bar{w}_{\beta 1})$ is degenerate, in which case the coefficients a_α^N $(\alpha = 1, \ldots, N)$ are non-unique. This can be seen by observing that u_2^N is the projection of u_2 on $\mathfrak{I}_{P1}^c \cap \mathfrak{I}_{P2}^c$, necessarily.

Conversely, if the sequence u_2^N converges, then one has

$$(u_2^N, \bar{w}_{\beta 1}) = \sum_{\alpha=1}^{N} a_\alpha^N (\bar{w}_{\alpha 1}, \bar{w}_{\beta 1}) = (w_{\beta 2}, \bar{V}_1), \quad \beta = 1, \ldots, N. \tag{12.29}$$

Taking the limit when $N \to \infty$,

$$(u_2', \bar{w}_{\beta 1}) = (w_{\beta 2}, \bar{V}_1), \quad \beta = 1, \ldots. \tag{12.30}$$

This shows

$$\langle A\hat{u}', \hat{w}_\beta \rangle = 0, \quad \hat{w}_\beta \in \mathfrak{B}.$$

Hence $\hat{u}' \in \hat{\mathfrak{I}}_P$ and (12.27) are straightforward.

References

1. Dunford, N. and J. T. Schwartz. *Linear operators*, Part II, Interscience, New York, 1963; Part III, 1971.
2. Arens, R. Operational calculus of linear relations. *Pacific J. Math.*, **11,** pp. 9–23, 1961.
3. Coddington, E. A. Self-adjoint subspace extensions of non-densely defined symmetric operators. *Advances in Math.*, **3**(4), pp. 309–331, 1974.
4. Coddington, E. A. Self-adjoint subspace extensions of non-densely defined symmetric operators. *Bull. Am. Math. Soc.*, **79,** pp. 712–715, 1973.
5. Coddington, E. A. Eigenfunction expansions for non-densely defined operators generated by symmetric ordinary differential expressions. *Bull. Am. Math. Soc.*, **79,** pp. 964–968, 1973.
6. Lee, S. J. Coordinatized adjoint subspaces in Hilbert spaces, with application to ordinary differential operators. *Proc. London Math. Soc.*, **41,** pp. 138–160, 1980.
7. Lee, S. J. On boundary conditions for linear differential operators. *J. London Math. Soc.*, **12**(2), pp. 447–454, 1976.
8. Lee, S. J. Operators generated by countably many differential operators. *J. Differential Equations*, **29**(3), pp. 453–466, 1978.
9. Lee, S. J. Perturbations of operators, with application to ordinary differential operators. *Indiana University Math. J.*, **28**(2), pp. 291–309, 1979.
10. Lee, S. J. Index and nonhomogeneous conditions for linear manifolds, in *Spectral Theory of Differential Operators*, Ed. I. W. Knowles and R. T. Lewis, North Holland, Amsterdam, pp. 289–293, 1981.
11. Lee, S. J. Boundary conditions for linear manifolds, I. *J. Math. Analysis and Applications*, **73,** pp. 366–380, 1980.
12. Chapman, G. R. and S. J. Lee. Boundary conditions for linear manifolds, II. *J. Math. Analysis and Applications*, **3**(2), pp. 351–363, 1980.
13. Herrera, I. and J. Bielak. Discussion of proceeding paper 9152: Variational formulation of dynamics of fluid-saturated porous elastic solids. *J. Eng. Mech. Division, ASCE*, **99**(5), pp. 1097–1098, 1973.
14. Herrera, I. and J. Bielak. A simplified version of Gurtin's variational principles. *Arch. Rational Mech. and Analysis*, **53**(2), pp. 131–149, 1974.
15. Herrera, I. A general formulation of variational principles. *Inst. Ingeniería, UNAM*, **E-10,** 1974.
16. Herrera, I. Métodos variacionales para aplicaciones en Ingeniería y Física. Parte 1. Formulación general de los métodos variacionales. *Inst. Ingeniería, UNAM*, **349,** 1975.
17. Herrera, I. and J. Bielak. Applications of dual principles for diffusion equations, *Inst. Ingeniería, UNAM*, **E-12,** 1975.
18. Herrera, I. and J. Bielak. Dual variational principles for diffusion equations. *Quarterly of Applied Math.* **34**(1), pp. 85–102, 1976.

19. Herrera, I. and J. Bielak. Comments on the paper by J. N. Reddy: Modified Gurtin's variational principles in the linear dyanmic theory of viscoelasticity. *Int. J. Solids and Structures,* **13,** pp. 377–378, 1977.
20. Herrera, I. and M. J. Sewell. Dual extremal principles for non-negative unsymmetric operators. *J. Inst. Maths Applications,* **21,** pp. 95–115, 1978.
21. Herrera, I. General variational principles applicable to the hybrid element method. *Proc. Nat. Academy of Sciences, USA,* **74**(7), pp. 2595–2597, 1977.
22. Herrera, I. Theory of connectivity for formally symmetric operators. *Proc. Nat. Academy of Sciences, USA,* **74**(11), pp. 4722–4725, 1977.
23. Herrera, I. and F. J. Sabina. Variational principles for superelement formulations of diffraction problems. *Comunicaciones Técnicas, IIMAS-UNAM, Serie Investigación,* **10**(188), 28, 1979.
24. Herrera, I. On the variational principles of mechanics. *Trends in Applications of Pure Mathematics to Mechanics, II.* Ed. H. Zorsky, Pitman, London, pp. 115–128, 1979.
25. Herrera, I. and F. J. Sabina. Connectivity as an alternative to boundary integral equations. Construction of bases. *Proc. Nat. Academy of Sciences, USA,* **75**(5), pp. 2059–2063, 1978.
26. Herrera, I. Theory of connectivity: A systematic formulation of boundary element methods. *Applied Math. Modelling,* **3**(2), pp. 151–156, 1979.
27. Sabina, F. J., I. Herrera and R. England. Theory of connectivity: Applications to scattering of seismic waves. I. SH wave motion. *Proc. Second Int. Conf. on Microzonation, San Francisco, USA, Nov.* 26/*Dec.* 1, 1979.
28. Herrera, I. Theory of connectivity: A unified approach to boundary methods. In *Variational Methods in the Mechanics of Solids,* Ed. S. Nemat-Nasser, Pergamon Press, Oxford and New York, pp. 77–82, 1980.
29. England, R., F. J. Sabina and I. Herrera. Scattering of SH waves by surface cavities of arbitrary shape using boundary methods. *Physics of the Earth and Planetary Interiors,* **21,** pp. 148–157, 1980.
30. Sabina, F. J., R. England and I. Herrera. Scattering of SH waves by obstacles in half space. *Earthquake Notes,* **49,** pp. 75–76, 1978.
31. Herrera, I. Variational principles for problems with linear constraints. Prescribed jumps and continuation type restrictions. *J. Inst. Maths Applications,* **25,** pp. 67–96, 1980.
32. Herrera, I. Boundary methods in flow problems. *Proc. Third Int. Conf. on Finite Elements in Flow Problems, Banff, Canada, 10–13 June, 1980,* **1,** pp. 30–42. (Invited general lecture.)
33. Herrera, I. Boundary methods in water resources. *Finite Elements in Water Resources,* Ed. S. Y. Wang, *et al.*, The University of Mississippi, pp. 58–71, 1980. (Invited general lecture.)
34. Herrera, I. Boundary methods. A criterion for completeness. *Proc. Nat. Academy of Sciences, USA,* **77**(8), pp. 4395–4398, 1980.
35. Herrera, I. An algebraic theory of boundary value problems. *KINAM,* **3**(2), pp. 161–230, 1981.
36. Gourgeon, H. and I. Herrera. Boundary methods. *C*-complete systems for the biharmonic equation. In *Boundary Element Methods,* Ed. C. A. Brebbia, Springer-Verlag, Berlin, pp. 431–441, 1981.
37. Herrera, I. and D. A. Spence. Framework for biorthogonal Fourier series. *Proc. Nat. Academy of Sciences, USA,* **78**(12), pp. 7240–7244, 1981.
38. Herrera, I. Boundary methods for fluids. In *Finite Elements in Fluids,* Vol.

IV, Ed. R. H. Gallagher, D. Norrie, J. T. Oden and O. C. Zienkiewicz, J. Wiley, New York, Chapter 19, pp. 403–432, 1982.
39. Herrera, I. and H. Gourgeon. Boundary methods. C-complete systems for Stokes problems. *Computer Methods in Applied Mechanics and Engineering*, **30,** pp. 225–241, 1982.
40. Sánchez-Sesma, F. J., I. Herrera and J. Avilés. A boundary method for elastic wave diffraction. Application to scattering of SH-waves by surface irregularities. *Bull. Seismol. Soc. Am.*, **72**(2), 473–490, 1982.
41. Herrera, I. Boundary methods. Development of complete systems of solutions. In *Finite Element Flow Analysis*, Ed. T. Kawai, University of Tokyo Press, pp. 897–906, 1982. (Invited lecture.)
42. Alduncin, G. and I. Herrera. Solution of free boundary problems using C-complete systems. In *Boundary Element Methods in Engineering*, Ed. C. A. Brebbia, Springer-Verlag, Berlin, pp. 34–42, 1982.
43. Alduncin, G. and I. Herrera. Contribution to free boundary problems using boundary elements. Trefftz approach. *Cómunicaciones Técnicas, IIMAS-UNAM,* **331,** 1983, *Computer Methods in Applied Mechanics,* **32,** 1984 (in press).
44. Herrera, I. Trefftz method, chapter of *Boundary Element Methods*, Vol. 3, Ed. C. A. Brebbia, J. Wiley, New York, 1983.
45. Abraham, R. and J. E. Marsden, *Foundations of Mechanics.* Benjamin Cummings Publishing, Menlo Park, Calif., pp. 159–187, 1978.
46. Weinstein, A. Symplectic manifolds and their Lagrangian submanifolds. *Advances in Math.* (6), pp. 329–346, 1971.
47. Gurtin, M. E. Variational principles for linear initial value problems. *Quart. Appl. Math.*, **22,** pp. 252–256, 1964.
48. Gurtin, M. E. Variational principles for linear elastodynamics. *Arch. Rational Mech. Anal.*, **16,** pp. 34–50, 1964.
49. Mikhlin, S. G. *Variational Methods in Mathematical Physics.* Pergamon Press, Oxford, 1964.
50. Tonti, E. A systematic approach to the search for variational principles. *Int. Conf. on Variational Methods in Engineering, Southampton University*, Ed. C. A. Brebbia, Dept Civil Engineering, Southampton University, pp. 1.1–1.12, 1972.
51. Oden, J. T. and J. N. Reddy. *Variational Methods in Theoretical Mechanics.* Springer-Verlag, Berlin, Heidelberg, New York, 1976.
52. Prager, W. Variational principles of linear elastodynamics for discontinuous displacements, strains and stresses. In *Recent Progress in Applied Mechanics*, the *Folke-Adqvist Volume*, Ed. B. Brogerg, J. Hult, and F. Niordson, Almqvist and Wiksell, Stockholm, pp. 463–474, 1967.
53. Nemat-Nasser, S. General variational principles in non-linear and linear elasticity with applications. In *Mechanics Today*, **1,** pp. 214–261, 1972.
54. Nemat-Nasser, S. On variational methods in finite and incremental elastic deformation problems with discontinuous fields. *Quart. Appl. Math.*, **30,** pp. 143–156, 1972.
55. Mei, C. C. and H. S. Chen. A hybrid element method for steady linearized free surface flows. *Int. J. Num. Meth. Eng.*, **10,** pp. 1153–1175, 1976.
56. Zienkiewicz, O. C., D. W. Kelly and P. Bettess. The coupling of the finite element method and boundary solution procedures. *Int. J. Num. Meth. Eng.*, **11,** pp. 355–377, 1977.

57. Zienkiewicz, O. C. *The Finite Element Method in Engineering Science.* McGraw-Hill, New York, 1977.
58. Brebbia, C. A. *The Boundary Element Method for Engineers.* Pentech Press, London, 1978.
59. Brebbia, C. A. *Boundary Element Methods*, Ed. C. A. Brebbia, Springer-Verlag, Berlin, Heidelberg, New York, Vols 3 and 4, 1981 and 1982.
60. Bergman, S. *Integral Operators in the Theory of Linear Partial Differential Equations.* Springer-Verlag, Berlin, 1969.
61. Colton, D. *Solution of Boundary Value Problems by the Method of Integral Operators.* Research Notes in Mathematics, Pitman, London, 1976.
62. Gilbert, R. P. Constructive methods for elliptic equations. *Springer-Verlag Lecture Note Series*, Berlin, **365,** 1974.
63. Henrici, P. Complete systems of solutions for a class of singular elliptic partial differential equations. In *Boundary Problems in Differential Equations.* University of Wisconsin Press, Madison, 1960.
64. Vekua, I. N. *New Methods for Solving Elliptic Equations.* Wiley, New York, 1967.
65. Colton, D. and W. Watzlawek. Complete families of solutions to the heat equation and generalized heat equation in $R^{n'}$. *J. Differential Equations,* **25**(1), pp. 96–107, 1977.
66. Rektorys, K. *Survey of Applicable Mathematics,* Iliffe Books, London, 1969.
67. Trefftz, E. Ein Gegenstruck zum ritzs'schen Verfohren. *Proc. Second Int. Congress Appl. Mech., Zurich, 1926.*
68. Colton, D. The approximation of solutions to initial boundary value problems for parabolic equations in one space variable. *Quart. Appl. Math.,* **34**(4), pp. 377–386, 1976.
69. Rosenblueth, E. Personal communication, 1983.
70. Mikhlin, S. G. *Variational Methods in Mathematical Physics.* Pergamon Press, Oxford, 1964.
71. Rektorys, K. *Variational Methods in Mathematics, Science and Engineering.* D. Reidel, Hingham, Mass., 1977.
72. Kupradze, V. D., T. G. Gegelia, M. O. Basheleischvili and T. V. Burchuladze. *Three-dimensional Problems of the Mathematical Theory of Elasticity and Thermoelasticity.* North Holland, Amsterdam, 1979.
73. Amerio, L. Sul calculo delle autosoluzioni dei problemi al contorno per le equazioni lineari del secondo ordine di tipo ellitico. *Rend. Acc. Lincei,* **1,** pp. 352–359 and 505–509, 1946.
74. Fichera, G. Teoremi di completezza sulla frontiera di un dominio per taluni sistema di funzioni. *Ann. Mat. Pura e Appl.,* **27,** pp. 1–28, 1948.
75. Kupradze, V. D. On the approximate solution of problems in mathematical physics. *Russian Math. Surveys,* **22**(2), pp. 50–108, 1967 (*Uspehi Mat. Nauk,* **22**(2), pp. 59–107, 1967).
76. Picone, M. Nuovi metodi risolutivi per i problemi d'integrazione delle equazioni lineari a derivati parziali e nuova applicazione della transformate multipla di Laplace nel Caso delle equazioni a coefficienti constanti. *Atti. Acc. Sc. Torino,* **76,** pp. 413–426, 1940.
77. Colton, D. Complete families of solutions for parabolic equations with analytic coefficients. *SIAM J. Math. Anal.,* **6,** pp. 937–947, 1975.
78. Bates, R. H. T. Analytic constraints on electro-magnetic field computations. *IEEE Trans. on Microwave Theory of Techniques,* **23,** pp. 605–623, 1975.

79. Millar, R. F. The Rayleigh hypothesis and a related least-squares solution to scattering problems for periodic surfaces and other scatterers. *Radio Science*, **8,** pp. 785–796, 1973.
80. Millar, R. F. On the completeness of sets of solutions to the Helmholtz equation. *IMA J. Appl. Math.*, **30,** 27–38, 1983.
81. Oliveira, E. R. Plane stress analysis by a general integral method. *J. Eng. Mech. Div. ASCE*, **94,** pp. 79–101, 1968.
82. Lions, J. L. and E. Magenes. *Non-homogeneous Boundary Value Problems and Applications*. Springer-Verlag, New York, Heidelberg, Berlin (3 volumes), 1972.
83. Oden, J. T. and J. N. Reddy, *An Introduction to the Mathematical Theory of Finite Elements*. Pure and Applied Mathematics Series, J. Wiley, New York, London, Sydney, Toronto, 1976.
84. Miranda, C. *Partial Differential Equations of Elliptic Type*, 2nd edition. Springer, New York, 1970. (Translation of *Equazioni alle Derivate Parziali di Tipo Ellitico*, 1955.)
85. Joseph, D. D. A new separation of variables theory for problems of Stokes flow and elasticity. In *Trends in Applications of Pure Mathematics to Mechanics, II*. Ed. H. Zorski, Pitman, London, 1979.
86. Gregory, R. D. Green's functions, bi-linear forms, and completeness of the eigenfunctions for the elastostatic strip and wedge. *J. Elasticity*, **9**(3), 283–310, 1979.
87. Gregory, R. D. The semi-infinite strip $x \geq 0$, $-1 \leq y \leq 1$; completeness of the Papkovich–Fadle eigenfunctions when $\phi_{xx}(0, y)$, $\phi_{yy}(0, y)$ are prescribed. *J. Elasticity*, **10**(1), 57–80, 1980.
88. Smith, R. C. T. The bending of a semi-infinite strip. *Austr. J. Sci. Res.*, **5,** 227, 1952.
89. Herrera, I. On a method to obtain a Green's function for a multilayered half-space. *Bull. Seism. Soc. of Am.*, **54**(4), pp. 1087–1096, 1964.
90. Alsop, L. E. An orthonormality relation for elastic body waves. *Bull. Seism. Soc. of Am.*, **58**(6), pp. 1949–1954, 1968.
91. Malichewsky, P. Surface waves in media having lateral inhomogeneities. *Pure and Applied Geophysics*, **114,** 833–843, 1976.
92. Zamansky, M. *Linear Algebra and Analysis*. Van Nostrand, Amsterdam, 1969.
93. Birkhoff, G. and S. MacLane. *A Survey of Modern Algebra*. Macmillan, New York, 1953.
94. Courant, R. and D. Hilbert. *Methods of Mathematical Physics*, Volume 2, *Partial Differential Equations*. Interscience. (J. Wiley), New York, London, 1962.
95. Kellog, O. D. *Foundations of Potential Theory*. Dover Publications, New York, 1953.
96. Christiansen, S. On Kupradze's functional equations for plane harmonic problems. In *Function Theoretic Methods in Differential Equations*, Ed. R. P. Gilbert and R. J. Weinacht, Research Notes in Mathematics No. 8, Pitman, London, pp. 205–243, 1976.
97. Mei, C. C. and H. S. Chen. A hybrid element method for steady linearized free surface flows. *Int. J. Num. Meth. Eng.*, **10,** pp. 1153–1175, 1976.
98. Gurtin, M. E. The linear theory of elasticity. In *Encyclopedia of Physics*, VIa/2, Springer-Verlag, Berlin, pp. 1–295, 1972.

99. Temam, R. *Navier–Stokes Equations: Theory and Numerical Analysis.* North-Holland, Amsterdam, 1977.
100. Glowinski, R. and O. Pironneau. On numerical methods for the Stokes problem. Chapter 13 of *Energy Methods in Finite Element Analysis,* Ed. R. Glowinski, E. Y. Rodin and O. C. Zienkiewicz, J. Wiley, New York, 1978.
101. Aleksidze, M. A. Series in nonorthogonal systems of functions. *USSR Computational Math. and Math. Physics,* **8**(5), pp. 40–68, 1968.
102. Halmos, P. R. *Introduction to Hilbert Space and the Theory of Spectral Multiplicity.* Chelsea Publ., New York, 1951.
103. Yosida, K. *Functional Analysis.* Springer-Verlag, Berlin and Heidelberg, 1971.
104. Kupradze, V. D. and J. A. Aleksidze. The method of functional equations for the approximate solution of certain boundary value problems. *USSR Computational Math. and Math. Physics,* **4**(4), pp. 82–126, 1964.
105. Jackson, J. D. *Classical Electrodynamics.* Wiley, New York, pp. 65, 69, 86 and 541, 1962.
106. Morse, P. M. and H. Feshbach. *Methods of Theoretical Physics,* Vol. II. McGraw-Hill, New York, p. 827, 1953.
107. Whittaker, E. T. and G. N. Watson. *A Course of Modern Analysis.* Cambridge University Press, 1958.
108. Herrera, I. and J. Bielak. Soil-structure interaction as a diffraction problem, *Proc. VI World Conference on Earthquake Eng.,* pp. 4.19–4.24, New Delhi, 1977.